THE SCIENCE OF THE COUPLE:

The Ontario Symposium Volume 12

edited by

Lorne Campbell
Jennifer G. La Guardia
James M. Olson
Mark P. Zanna

LONDON AND NEW YORK

First published 2002 by Psychology Press

Published 2017 by Routledge
2 Park Square, Milton Park, Abingdon, Oxon OX14 4RN
711 Third Avenue, New York, NY 10017, USA

First issued in paperback 2017

Routledge is an imprint of the Taylor & Francis Group, an informa business

ISBN 13: 978-1-138-11078-6 (pbk)
ISBN 13: 978-1-84872-979-7 (hbk)

Visit the Taylor & Francis Web site at
http://www.taylorandfrancis.com

Contents

List of Contributors

Lane Beckes
Department of Psychology
University of Virginia
Charlottesville, Virginia

Lorne Campbell
Department of Psychology
University of Western Ontario
Ontario, Canada

James A. Coan
Department of Psychology and the Neuroscience Graduate Program
University of Virginia
Charlottesville, Virginia

Eli J. Finkel
Department of Psychology
Northwestern University
Evanston, Illinois

Gráinne M. Fitzsimons
Department of Psychology
University of Waterloo
Ontario, Canada

Shelly L. Gable
Department of Psychology
University of California–Santa Barbara
Santa Barbara, California

John G. Holmes
Department of Psychology
University of Waterloo
Ontario, Canada

Jennifer G. La Guardia
Healthy Living Center
Center for Community Health
University of Rochester
Rochester, New York

Lisa Linardatos
Department of Psychology
McGill University
Quebec, Canada

Timothy J. Loving
Department of Human Development and Family Sciences
The University of Texas at Austin
Austin, Texas

John E. Lydon
Department of Psychology
McGill University
Quebec, Canada

Mario Mikulincer
School of Psychology
Interdisciplinary Center (IDC) Herzliya
Tel Aviv, Israel

Sandra L. Murray
Department of Psychology
University at Buffalo
State University of New York
Buffalo, New York

Rebecca T. Pinkus
Department of Psychology
University of Western Sydney
New South Wales, Australia

Harris Rubin
Department of Psychology
University of Western Ontario
Ontario, Canada

David A. Sbarra
Department of Psychology
University of Arizona
Tuscon, Arizona

Phillip R. Shaver
Department of Psychology
University of California–Davis
Davis, California

Jeffry A. Simpson
Department of Psychology
University of Minnesota
Minneapolis, Minnesota

SiSi Tran
Department of Psychology
University of Toronto Scarborough
Ontario, Canada

Brittany L. Wright
Department of Human Development and Family Sciences
The University of Texas at Austin
Austin, Texas

Preface

The Twelfth Ontario Symposium on Personality and Social Psychology was held at the University of Western Ontario in London, Ontario, Canada, August 20–21, 2009. A total of 13 internationally renowned scholars presented their recent theoretical and empirical work on the topic of the symposium, "The Science of the Couple." The presentations were very thought provoking and there was a lively discussion of ideas. In addition to the invited presenters, approximately 100 faculty and graduate students from near and abroad were in attendance. Additionally, over 50 individuals presented posters highlighting their most recent research on close relationship processes. As a lasting legacy of this symposium, the current volume contains updated and edited versions of 12 of the papers that were presented.

How did we come to focus on the science of the couple? Although love and relationships have been focal points for poets and philosophers for thousands of years, these topics have not traditionally been the focus of empirical research. As a result, very little was known about how couples maintained happiness and satisfaction in their relationships or how relationships deteriorated and ultimately ended in separation or divorce. Only two decades ago, Harold Kelley and his colleagues (1983) called for a comprehensive study of love and close relationships, and since that time relationship research has blossomed as a field, with studies on close relationships routinely published in top journals such as the *Journal of Personality and Social Psychology*, *Personality and Social Psychology Bulletin*, and the *Journal of Marriage and Family*. Two journals (*Journal of Social and Personal Relationships* and *Personal Relationships*) were created solely for the purpose of publishing studies in close relationships. We therefore felt that this symposium and this accompanying edited volume were timely given the tremendous growth in the relationships field.

To facilitate this goal, invited speakers represented social, personality, neurobiology, health, and clinical perspectives on couple functioning. Additionally, the research presented highlighted the use of survey, experimental, implicit, and longitudinal methods as well as specialized techniques employed in neuroscience, psychophysiology, and psychoneuroimmunology in the study of couple-level processes. The broader aim of the symposium and this edited volume was to develop a dialogue about ways these theories and methods may converge to provide a deeper, holistic model of couples' processes and functioning.

The first chapter in the book, by Lorne Campbell and Harris Rubin, focuses on research designs that should optimally model dyadic processes. Using myriad methods of data collection as well as samples of individuals and couples, they provide examples of research that excel at modeling the interconnections between partner members and provide suggestions for how best to study relationship processes. In Chapter 2, Phillip Shaver and Mario Mikulincer provide an in-depth analysis of anxious attachment in adult romantic relationships and link anxious attachment to specific motivational patterns in close relationships. A social cognitive approach to relationship processes is provided by Gráinne Fitzsimons and Eli Finkel in Chapter 3. They discuss how individuals can "outsource effort" to their close partners, potentially resulting in reduced individual effort being expended to achieve personal goals. Shelly Gable stresses the importance of approach and avoidance relationship goals in Chapter 4. She argues that relationship researchers need to consider the motivating role of potential incentives in relationships instead of focusing primarily on how individuals deal with potential threat. In Chapter 5, Lane Beckes and James Coan discuss the physiological regulation of emotion in relationships from the perspective of social baseline theory. They argue that close relationships have been of great importance to humans over evolutionary history and that coping with stressful circumstances is more successful when with close others than alone.

In Chapter 6, SiSi Tran and Jeffry Simpson adopt a dyadic approach to understand the role of constructive emotion regulation for relationship well-being. They discuss how ideas from both attachment and interdependence theories can be combined to optimally predict relationship outcomes. John Lydon and Lisa Linardatos focus on the motivational bases of commitment to relationships and whether different motives differentially predict personal and relational well-being in Chapter 7. In a similar vein, in Chapter 8, Sandra Murray and Rebecca Pinkus suggest the presence of a “smart relationship unconscious” that underlies motivated commitment to relationships. Timothy Loving and Brittany Wright discuss the origins of research on stress in Chapter 9 and propose that researchers have to date ignored the role of eustress (i.e., a positive form of stress, such as falling in love) in close relationships. They suggest that studying eustress is essential to a greater understanding of the mind–body connection. In Chapter 10, Jennifer La Guardia addresses the emotional challenges facing couples following one partner’s health crisis and discusses how couples’ interactions after such an event are critical to the subsequent health of both partners and the functioning of the relationship. David Sbarra provides a compelling review of the literature on marital dissolution and health outcomes in Chapter 11 and suggests that much more research is needed to understand the impact of marital dissolution on individuals beyond documenting that it is an unpleasant experience. Last, in Chapter 12, John Holmes provides an integrative view of the research presented in the prior chapters and offers suggestions for new directions for interdisciplinary collaborations in the field in the future.

As in previous years, the Social Sciences and Humanities Research Council of Canada, whose continuing support has been the backbone of the series, provided primary financial support for the 12th Ontario Symposium. We are also deeply indebted to the Faculty of Social Science at the University of Western Ontario for its financial support. We would like to thank the many graduate students at the University of Western Ontario who voluntarily provided assistance in conducting this conference. Finally, we thank Paul Dukes and the editorial team at Psychology Press for their support and editorial guidance.

We believe that the chapters appearing in this edited volume provide a solid foundation for the future direction of research on couple-level processes. We hope that you enjoy them as much as we have.

Lorne Campbell
Jennifer G. La Guardia
James M. Olson
Mark P. Zanna

Table P.1 History of the Ontario Symposium

Date	Topic	Author(s) and Year	Location
August 1978	Social cognition	Higgins, Herman, and Zanna (1981)	University of Western Ontario
October 1979	Variability and consistency in social behavior	Zanna, Higgins, and Herman (1982)	University of Waterloo
May 1981	Social psychology of physical appearance	Herman, Zanna, and Higgins (1986)	University of Toronto
October 1983	Relative deprivation and social comparison processes	Olson, Herman, and Zanna (1986)	University of Western Ontario
August 1984	Social influence processes	Zanna, Olson, and Herman (1987)	University of Waterloo
June 1988	Self-inference processes	Olson and Zanna (1990)	University of Western Ontario
June 1991	Prejudice	Zanna and Olson (1994)	University of Waterloo
August 1993	Psychology of values	Seligman, Olson, and Zanna (1996)	University of Western Ontario
May 2000	Motivated social perception	Spencer, Fein, Zanna, and Olson (2003)	University of Waterloo
June 2002	Culture and social behavior	Sorrentino, Cohen, Olson, and Zanna (2005)	University of Western Ontario
August 2007	Psychology of justice and legitimacy	Bobocel, Kay, Zanna, and Olson (2010)	University of Waterloo
August 2009	The science of the couple	Campbell, La Guardia, Olson, and Zanna (2012)	University of Western Ontario

References

Higgins, E.T., Herman, C.P., & Zanna, M.P. (1981). *Social cognition: The Ontario symposium* (Vol. 1). Hillsdale, NJ: Erlbaum.

Zanna, M.P., Higgins, E.T., & Herman, C.P. (1982). *Consistency in social behavior: The Ontario symposium* (Vol. 2). Hillsdale, NJ: Erlbaum.

Herman, C.P., Zanna, M.P., & Higgins, E.T. (1986). *Physical appearance, stigma, and social behavior: The Ontario symposium* (Vol. 3). Hillsdale, NJ: Erlbaum.

Olson, J.M., Herman, C.P., & Zanna, M.P. (1986). *Relative deprivation and social comparison: The Ontario symposium* (Vol. 4). Hillsdale, NJ: Erlbaum.

Zanna, M.P., Olson, J.M., & Herman, C.P. (1987). *Social influence: The Ontario symposium* (Vol. 5). Hillsdale, NJ: Erlbaum.

Olson, J.M., Zanna, M.P., & Herman, C.P. (1990). *Self-inference processes: The Ontario symposium* (Vol. 6). Hillsdale, NJ: Erlbaum.

Zanna, M.P., & Olson, J.M. (1994). *The psychology of prejudice: The Ontario symposium* (Vol. 7). Hillsdale, NJ: Erlbaum.

Seligman, C., Olson, J.M., & Zanna, M.P. (1996). *The psychology of values: The Ontario symposium* (Vol. 8). Hillsdale, NJ: Erlbaum.

Spencer, S.J., Fein, S., Zanna, M.P., & Olson, J.M. (2003). *Motivated social perception: The Ontario symposium* (Vol. 9). Hillsdale, NJ: Erlbaum.

Sorrentino, R.M., Cohen, D., Olson, J.M., & Zanna, M.P. (2005). *Culture and social behavior: The Ontario symposium* (Vol. 10). Hillsdale, NJ: Erlbaum.

Bobocel, D.R., Kay, A.C., Zanna, M.P., & Olson, J.M. (2010). *The psychology of justice and legitimacy: The Ontario symposium* (Vol. 11). New York: Psychology Press.

Campbell, L., LaGuardia, J., Olson, J.M., & Zanna, M.P. (2012). *The science of the couple: The Ontario symposium* (Vol. 12). New York: Psychology Press.

CHAPTER 1

Modeling Dyadic Processes

LORNE CAMPBELL
University of Western Ontario

HARRIS RUBIN
University of Western Ontario

> Reflecting psychology's individualistic orientation, virtually all of our methodologies and statistics are predicated on the individual as the unit of analysis. As a consequence, relationship scholars often find themselves jerry-rigging old methodologies and statistics to accommodate the dyadic unit of analysis, but some, such as David Kenny and his associates (e.g., Kenny & La Voie, 1984), are creatively creating new ones.
>
> **Berscheid (1999, p. 261)**

The empirical study of close relationships is inherently complex. A relationship involves at least two people (e.g., spouses, parent and child, two close friends), which means that researchers need to move beyond a focus on the individual to a focus on the interconnections between individuals. Additionally, when data from two or more relationship members are collected, many statistical challenges arise. Berscheid (1999) eloquently acknowledged these challenges and encouraged relationship researchers to develop methodologies and data analytic approaches to optimally model dyadic processes.

Fortunately, for current relationship researchers, the development of dyadic data analytic techniques and the publication of clear and helpful guides on how to practically implement these techniques has grown considerably as interest in the scientific study of close relationships has risen. A researcher with an interest in collecting dyadic data can turn to one of several excellent resources to learn how to properly take into consideration the nonindependence that exists between dyad members' scores on study variables in myriad ways (e.g., Boker & Laurenceau, 2006, 2007; Gonzalez & Griffin, 1997, 1999, 2002; Griffin & Gonzalez, 1995; Gottman, Murray, Swanson, Tyson, & Swanson, 2002; Kashy, Campbell, & Harris, 2006; Kashy & Kenny, 2000; Kashy & Snyder, 1995; Kenny, 1994, 1995, 1996; Kenny & Cook, 1999; Kenny, Kashy, & Bolger, 1998; Kurdek, 2003; West, Popp, & Kenny, 2008) or to obtain the necessary computer syntax for a variety of statistical software packages to assist in data analysis (e.g., Campbell & Kashy, 2002; Kenny, Kashy, & Cook, 2006). The result of this gradual refinement of theory and application has armed relationship researchers with a set of powerful and complex data analytic tools.

Given that these statistical techniques are very well documented, we will not flog the proverbial dead horse by providing yet another discussion of how to statistically model dyadic processes in close relationships. Instead, we take a step back and focus on designing research that collects data in a manner that optimally models dyadic processes. A critical first step in this discussion, however, is to define what we mean by *dyadic processes*. To do this, we turn to theoretical work that provides a compelling definition of close relationships as well as concrete suggestions for the study of dyadic processes.

Defining Close Relationships

Berscheid (1999) observed that relationships are akin to great forces of nature such as gravity, electricity, and the four winds in that they are powerful but ultimately invisible; we are aware of their existence only by observing their effects. A relationship can be discerned, therefore, by observing only the interconnections between two people and not simply the properties of each individual. As such, relationship science is generally concerned with developing a nomological network of laws regarding individuals' interactions with each other, or how people influence each other over time. Levinger (1980) echoed this sentiment when he asked, "Does the establishment of long-term relationships generate special products and investments that are connoted by a growing interpersonal intersection?" (p. 513). A unique aspect of relationship science, therefore, is that it transcends the study of the individual and focuses on the unique pattern of interconnections that evolve *between* individuals.

Perhaps the most thorough and influential theoretical discussion of relationships was undertaken by Kelley and colleagues in their 1983 book titled *Close Relationships*. Kelley et al. identified the defining elements of a close relationship, noting that it "is one of strong, frequent, and diverse interdependence that lasts over a considerable period of time" (1983, p. 38). Within this interdependence framework, behavioral, cognitive, and affective events for both partners lead to a perpetual feedback loop of mutual influence across time. As such, the essence of this conception of a relationship centers around how one's own subjective relationship experience is partly based on one's partner's thoughts, feelings, and behaviors. A quintessential feature of close relationships, therefore, is that partners' thoughts, feelings, and behaviors are causally connected.

Consistent with this definition of close relationships, virtually all major models of relationship processes incorporate the concept of interdependence, including theories of attachment (Bowlby, 1969, 1973, 1980), commitment (Rusbult, 1980), equity (Messick & Crook, 1983; Walster, Walster, & Berscheid, 1978), interdependence (Kelley & Thibaut, 1978; Rusbult, 1983; Thibaut & Kelley, 1959), perceived partner responsiveness (e.g., Reis, 2007; Reis, Clark, & Holmes, 2004; see also Reis & Shaver, 1988), self-expansion (Aron, Aron, Tudor, & Nelson, 1991), and trust (Rempel, Holmes, & Zanna, 1985). Mutual influence is also germane to other types of important dyadic relationships (e.g., friendships and parent–child relationships). Theories of close relationships acknowledge that the relationship exists primarily in the interactive space between individuals and not simply within individuals.

One important implication of this definition of close relationships is that modeling dyadic processes, both in terms of how studies are designed and how data are analyzed, requires a focus on the interconnections between at least two people. As already stated, the relationship exists only because of these interconnections, and the patterns of these interconnections define the quality of the relationship. Within this framework, the most direct route to studying and understanding relationship processes should involve the collection of data from both members of the relationship given that the interconnections that define the relationship can be optimally assessed by observing each member of the relationship.

One goal of this chapter is to discuss various research methodologies that are optimally suited to test relationship processes in light of the aforementioned definition of a close relationship. Prior to doing so, however, we first want to get a sense of the type of data collected by relationship researchers to test their hypotheses. As a field, do we focus relatively more attention on the individual compared with the "relationship"? To do this, we turn to a fairly simple survey of the close relationship literature reported by Kashy, Campbell, and Harris (2006).

Type of Data Gathered in the Relationship Sciences

Kashy et al. (2006) wanted to determine the frequency of use of the new statistical strategies developed specifically for dyadic data and if the adoption of these techniques had increased over time. To accomplish this goal, they surveyed research focusing on close relationships from five prominent journals that publish relationships research: *Personal Relationships, Journal of Social and Personal Relationships, Journal of Personality and Social Psychology—Interpersonal Relations and Group Processes, Personality and Social Psychology Bulletin,* and *Journal of Marriage and Family.* All research publications within the close relationships domain from these journals for the years 1994 (n = 157) and 2002 (n = 181) were examined, with a specific focus on two aspects of the research: (1) the unit from which the data were collected; and (2) the data analytic strategy applied.

For each study in the sample, it was determined whether the study collected data from one individual in a given relationship, from both members of a dyadic relationship, or from multiple family members or members of other relationship groups (e.g., friendship groups where group size is greater than two). The individual data code therefore included studies in which an individual provided data at one time point about one relationship as well as studies in which an individual provided data at many time points about one or many relationships. The key element of the individual code is only that one individual's perspective on the relationships in question was obtained. When the study collected data from more than one person (e.g., both marital partners; parent and child), the data analytic strategy was also coded.

A very interesting finding emerged from this simple survey. Less than one-quarter of the relationship-oriented research published in the journals sampled involved dyadic data in both 1994 (22.9%) and 2002 (24.3%). Additionally, less than 5% of data were generated by families or other small groups (4.5% in 1994 and 2.2% in 2002). Consequently, the majority of research published on relationship processes relied on the perspective of only one person in the relationship (72.6% in 1994 and 73.5% in 2002). Remarkably, these trends remained very stable across the two time points sampled. From this survey, one possible conclusion is that the causal connections that are believed to be the quintessential feature of a close relationship seem to be the target of a minority of relationship-based research, implying perhaps the presence of a disconnect between the theoretical definition of a close relationship and the scientific study of close relationships. This reality also indicates that the voluminous published resources for analyzing dyadic data cited earlier cater to a minority of relationship-based research.

There are many possible reasons for this pattern of data collection in the relationship science. For instance, it may stem from the increased difficulty (e.g., higher costs and time commitment) associated with collecting data from both members of a close relationship. It may also be the case that many researchers do not have the ability to recruit relationship partners to their studies.

One may also infer, however, that this pattern of data collection and hypothesis testing stems from some close relationship researchers being more concerned with the individual-level psychological processes of people involved in close relationships rather than psychological processes that operate at the dyadic level. Can research that on the surface overlooks the interdependence that is fundamental to close relationships or that seems to be focused on individual-level processes truly be classified as relationship research? Does this body of research genuinely advance our knowledge of relationship processes?

The answer to these questions is a qualified yes. In some circumstances, research designs that collect data from only one member of a relationship can do a respectable job of modeling dyadic processes as defined by Kelley et al. (1983). Specifically, research on individuals in relationships that focuses on the connections between relationship members can arguably be directly assessing relationship processes. To cite an example, Murray, Rose, Bellavia, Holmes, and Kusche (2002, Studies 1 and 2) used a clever set of manipulations to directly activate relationship-relevant partner knowledge in study participants, allowing them to observe how people respond to their partner, even in their partner's absence. In this research, participants were first led to realistically believe that their partner perceived a problem with their relationship. Following this manipulation, participants answered a number of questions tapping outcomes linked to the self, their partner, and their relationship. The results of these experiments showed that individuals with low self-esteem were more likely to derogate their partner and minimize felt closeness to their partner in the face of relationship threat compared with high-self-esteem participants. A third experiment using a dyadic sample to more realistically activate relationship threat yielded a very similar pattern of effects. Murray et al. were therefore successful in developing a research strategy that effectively assessed the interdependence that exists between romantic partners by activating relationship-relevant concerns in a realistic and believable manner. That the causal influence of interdependence is assessed in only one direction (i.e., the effects on partner B after activating thoughts of partner A) and at only one point in time of course illustrates important limitations of this approach.

An equally important question to ask is whether dyadic processes are always being addressed in research designs where data from both members of a relationship has been collected. We feel that the answer to this question is no. That is, in many cases the researcher has collected dyadic data, but the study tests hypotheses focused only on individual-level processes (see also Kashy et al., 2006). For example, in an excellent study of interpersonal perception and marital discord, Neff and Karney (2005) used data from a sample of newlyweds to test the hypothesis that individuals who viewed their partners as globally positive but who were accurate in their perceptions of more specific traits should be more satisfied with their marriage and less likely to divorce. Using analytic techniques that statistically controlled for the interdependence between spouses, they generated separate estimates for husbands and wives and their hypotheses were confirmed only for wives. Although longitudinal data were collected from both members of these newlywed couples, the study hypotheses focused primarily on individual-level phenomena and thus did not directly model dyadic processes.

Simply having dyadic data in hand is therefore not a sufficient condition for testing dyadic-level processes, and that data are collected from only one person of a close relationship does not necessarily mean that the ability to model dyadic processes, to at least some degree, is absent. Rather than being dependent on whether data are collected from both members of a relationship, effectively teasing apart the dyadic processes associated with the interdependence of a close relationship hinges on the type of research questions that are put forward and tested. The field of close relationships is rich in theory incorporating the concept of interdependence, implying that the bulk of close relationship research should be focusing on the interconnections between people instead of individual psychological processes of people that happen to be involved in a particular close relationship. In the opening quote from Berscheid (1999), she suggested that to achieve this goal our field needs to develop new statistical approaches for the analysis of both dyadic data and new methodologies for study of dyadic processes. The statistical tools have been, and continue to be, developed, and research designs that creatively model dyadic processes are increasingly being used by relationship researchers. We feel that more effort should be expended, however, on constructing and implementing research designs that more directly model dyadic and not simply individual psychological processes in close relationships. To facilitate this goal, in the next section we discuss different types of investigative methods available to researchers and highlight extant research that has effectively used these approaches to model dyadic processes.

Investigative Methods

As is the case in any research domain, the selection of a particular research design means accepting some limitations to take advantage of the strengths associated with that particular approach. To organize our discussion of how individual- and dyadic-level data can be leveraged to test dyadic processes, we use the research strategy classification system developed by Runkel and McGrath (1972; see also McGrath, 1982).

In this classification system, a given research strategy falls along two orthogonal dimensions: (1) the degree to which the strategy uses obtrusive versus unobtrusive procedures, and (2) the degree to which the strategy captures universal versus specific behavior systems. These dimensions are further divided into a circumplex with four quadrants. The selection of a strategy within one of these quadrants represents a trade-off between the mutually conflicting research goals of (1) generalizeability of results across populations, (2) the precision with which variables are measured, and (3) the realism of the context in which variables are measured. Organizing research strategies in this manner reveals a dilemma researchers must face—namely, that these research goals cannot be simultaneously maximized. Research designs that maximize precision of measurement, for example, sacrifice the realism of the context in which the study occurs. As such, to achieve all of these research goals, relationship researchers must test their theories using a combination of research strategies (i.e., use a multimethod approach; Brewer, 2000).

Researchers wishing to model close relationship processes are faced with the additional complication of whether to sample individuals or dyads. To illustrate our assertion that studying dyadic processes does not always require collecting data from both members of dyads, we use examples of research strategies across these quadrants that have successfully modeled dyadic processes by sampling either individuals or dyads.

Experiments

A hallmark of social psychology has been the field's ability to use experimental methods. Experimental research strategies are defined by a high degree of experimenter control that, in combination with random assignment, allows for the ability to make strong causal conclusions. Participants are typically randomly assigned to two or more conditions that are identical except for the critical manipulations. Any differences that emerge across conditions can thus be attributed to the manipulations. In the case of dyadic processes, an experimental strategy seeks to identify contextual variables that impact the thoughts, feelings, and behaviors of relationship members. Experiments therefore optimize the research goal of precision of measurement and in many cases do a good job of creating psychological realism for study participants.

On the face of it, the physical presence of a relationship partner seems like a necessary condition for identifying contextual factors that lead to theoretically important relationship outcomes. On the contrary, examples abound of the mere psychological presence of a relationship partner being sufficient to test dyadic hypotheses. For instance, using a series of experiments, Fitzsimons and Bargh (2003) found that priming individuals with various types of relationship partners (e.g., friends, parents) leads to the activation of a predictable pattern of interpersonal goals, which in turn leads to goal-relevant thoughts and behaviors. For example, priming individuals with a best friend was found to be linked to a greater reported intent to provide help when compared with when individuals were primed with a liked coworker. A separate study using a behavioral outcome variable tapping achievement motivation (i.e., number of words found in an anagram task) showed that individuals primed with their mother found more words than individuals primed with a friend.

The supraliminal and subliminal priming techniques used in these experiments illustrate how our mental representations of close others are composed of diverse components such as interpersonal goals that can operate at an automatic level. By attending to the features of these representations when designing their experimental manipulations, Fitzsimons and Bargh (2003) provide an excellent example of how to tap into processes that are dyadic in nature with a nondyadic sample.

Despite the ability to test dyadic hypotheses using data from individuals, having both members of a close relationship in the lab often allows for powerful manipulations based on more ecologically valid relationship situations. A recent example of a laboratory experiment using a dyadic sample tested the notion that people prefer to be viewed in both a verifying and enhancing manner from their romantic partners (Lackenbauer, Campbell, Rubin, Fletcher, & Troister, 2010). In this research, both members of the couple participated in the study together but in separate rooms. Each partner was asked to provide self- and partner evaluations on 10 interpersonal traits, and then they were informed that the computer would display a comparison of these ratings for each partner (i.e., participants would be able to view how they were perceived by their partner relative to their own self perceptions). The partner evaluations were manipulated, however, to result in feedback that varied in positive bias (high or low) and tracking accuracy (accurate or inaccurate). Results from dyadic analyses showed that participants responded positively to both positively biased and accurate feedback but particularly preferred feedback that was both accurate and positively biased. One strength of this experimental design is that the presence of both partners in the lab allowed for the manipulated feedback received from one's romantic partner to seem realistic, despite the relatively contrived social context (see also Campbell, 2005; Campbell, Lackenbauer & Muise, 2006; Murray et al., 2002).

Additionally, research by Simpson, Rholes, and Phillips (1996) combined an experimental and correlational approach. In this study they measured attachment orientations of each partner at an intake session and then videotaped couples discussing conflicts in the lab, manipulating the severity of conflicts that romantic partners were asked to discuss (minor or major problem). This research design allowed Simpson and colleagues to directly observe the interconnections between partners' behavior across the two experimentally contrived contexts. The behavior of partners during the conflict discussion was videotaped, and the results revealed that more anxiously attached women were observed to be more distressed by discussing major problems in their relationship, whereas more avoidantly attached men were observed to be less warm and supportive when discussing major relative to minor problems in their relationship.

With the advantages of precise measurement, random assignment, and causal conclusions from experiments come issues of low generalizability and ecological validity. The thoughts, feelings, and behaviors that comprise a close relationship naturally unfold without the obtrusiveness of researchers. As such, it is important that some close relationship theory testing occurs in more natural settings.

Field Research

The primary benefit of field research is that it maximizes mundane realism (i.e., the degree to which the measurement conditions resemble daily life). Field research is relatively underused by relationships researchers in social psychology. This is unfortunate because this sort of research has the ability to capture the operation of dyadic processes without the influence of unrealistic settings and obtrusive measurement. One example of field research that illustrates the positive characteristics of field work is a study by Fraley and Shaver (1998) in which they unobtrusively observed over 100 couples prior to boarding a flight in an airport. This study was able to show that many attachment behaviors exhibited in childhood have parallels to how adults behave within romantic relationships. For instance, couples who were separating at the airport sought more contact, were more caregiving, and engaged in more sexual touching when compared with couples who were preparing to board a flight together. Analyses using individual differences in adult attachment orientations showed attachment anxiety was linked to higher levels of distress in separating couples, but only for women. This sort of naturalistic observation proved to be an ideal tool for modeling dyadic processes within an attachment theory framework and has the capability of being used more widely to test a variety of hypotheses regarding relationship processes.

Although field research is able to capture people's thoughts, feelings, and behaviors in settings that naturally occur, these methods typically suffer from relatively low levels of measurement precision and control. For example, in Fraley and Shaver's (1998) field study, their conclusions are restricted to the individuals who happened to be traveling by plane rather than to a broader population of individuals. To achieve higher levels of generalizability, however, mundane realism must unfortunately be sacrificed.

Surveys

Survey research provides a means to obtain results that are highly generalizable but in return sacrifices mundane realism and precision of measurement. In one well-known and classic example, Hazan and Shaver (1987) surveyed a large sample of individuals to determine if the attachment classifications used in parent–child relationships could also be used to understand relationship processes between adults. In one study, local newspaper readers were asked to circle one of the three descriptions that best characterized their romantic relationship experiences (i.e., avoidant, ambivalent, or secure) and were then asked to answer some standard relationship quality questionnaires. A second study recruited undergraduate students to answer the same questions, and the results were consistent across samples. Overall, it was concluded that love can be conceptualized as an attachment process and that individual differences in romantic attachment styles were related to predicable relationship outcomes. Although this research was completely devoid of social context, it offered highly generalizable results that provided the basis for a massive program of research studying adult attachment.

An example of survey research that modeled dyadic processes using a dyadic sample investigated the role adult attachment orientations played in the transition to parenthood (Rholes, Simpson, Campbell, & Grich, 2001). By collecting survey data from husbands and wives in two waves (6 weeks prior to childbirth and 6 months postpartum), this study found that insecure attachment was negatively associated with martial satisfaction and functioning during the transition to parenthood. Commensurate with Kelley et al.'s (1983) conception of the close relationship construct, the hypotheses in this research were primarily focused on dyadic processes such as perceptions of spousal support, support-seeking behaviors, and how these attachment-relevant thoughts and behaviors varied over time. Several interesting longitudinal and sex-specific partner effects emerged, such as anxiously attached wives who perceived their husbands as less supportive at time 1 had husbands who reported being less satisfied at time 2. The benefit of having data from both partners is that this mutual influence between partners could be fruitfully investigated.

Formal Theory

Formal theory attempts to specify sets of propositions that describe universal behavior systems that are generalizable to broader populations. Within the field of relationship research, there exists substantial variability in the degree to which theories are focused on individual-level or relationship-level processes. For example, one perspective that is more focused on psychological processes occurring at the level of the individual is Baumeister and Bratslavsky's (1999) model of relationship intimacy and passion. Their theory holds that, rather than being directly related, passion's association with intimacy is best understood as being linked to changes in intimacy over time. Within this framework, when intimacy shows relatively large and rapid increases, levels of passion should be high. When intimacy remains unchanged over time, levels of passionate experience should be low. Although this theory directly addresses thoughts, feelings, and behaviors that occur within a relationship context, its main tenets are focused on individual-level changes in one's perceived intimacy and passion.

The model of perceived partner responsiveness (PPR; Reis, 2007; Reis, Clark, & Holmes, 2004) provides an example of a relationships theory that directly models dyad-level processes. The PPR construct rests on overlapping elements of beliefs about the partner's understanding of aspects of the self, beliefs about how much a partner values the self, and beliefs about how supportive a partner is toward oneself. Thus, this model is interested in the interdependence between the self and partner in dyadic context. As such, research designs applying the PPR construct model dyadic processes in a manner consonant with Kelley et al.'s (1983) conception of a close relationship.

Because most research is designed to test hypotheses derived from theory, how focused a particular close relationships theory is on individual- versus dyad-level processes is critical for how well its research program models interdependence. Granted, researchers have the ability to derive dyadic hypotheses with individualistic theories, but the form of a theory may constrain one's ability to model dyadic processes.

Summary and Conclusions

The central theme of this chapter is that true relationship research models dyadic processes by directly focusing on the interconnections between at least two people. A variety of research methods can be used to achieve this goal, as the examples we discussed suggest. Indeed, a multimethod approach is critical to any field of research (e.g., Brewer, 2000) and should be a priority for relationship researchers. In surveying the literature we were happy to see that relationship scholars incorporate survey, experimental, and field research methods to test their hypotheses. The diversity of research methods employed represents an important strength of the field.

Modeling dyadic processes can also be achieved by collecting data from individuals as well as from all relationship members. Some creative research has sought to activate relationship interdependence to observe how individuals respond on subsequent tasks, thereby attempting to assess the interconnections between individuals and not simply asking people how they perceive their partners and relationships. Reiterating an earlier point, it is also the case that the collection of dyadic data does not necessarily mean that the researcher is investigating dyadic processes. That being said, how relationship partners influence each other in the relationship can be observed more directly when data are obtained from both (or more) partners. At present, however, less than a quarter of relationship-focused research obtains these types of data on a routine basis. This reality implies that one fruitful avenue for future relationship research is to develop methods and procedures to test relationship processes by collecting data from two or more individuals involved in the relationship of interest. Focusing on patterns of interpersonal influence over time is also essential to broaden our knowledge of relationship processes. The statistical procedures exist for the analysis of these types of data, and we anticipate that collecting data from more than one individual in the relationship, and over time, will grow in the future.

The Science of the Couple

The chapters in this volume reflect the theme of this chapter in that they all highlight the interdependent nature of close relationships, particularly romantic relationships, but do so by using diverse samples and methods. The ideas and research presented in these chapters demonstrate that how well we are able to model dyadic processes hinges on how we as researchers define close relationships. The researchers contributing this volume embrace Berscheid's (1999) wise observation that relationships can only ever be observed indirectly based on their effects, meaning that our hypotheses and methods need to focus on the interactive space between individuals.

References

Aron, A., Aron, E. N., Tudor, M., & Nelson, G. (1991). Close relationships as including other in the self. *Journal of Personality and Social Psychology, 60*, 241–253.

Baumeister, R. F., & Bratslavsky, E. (1999). Passion, intimacy, and time: Passionate love as a function of change in intimacy. *Personality and Social Psychology Review, 3*, 49–67.

Berscheid, E. (1999). The greening of relationship science. *American Psychologist, 54*, 260–266.

Boker, S. M., & Laurenceau, J. P. (2006). Dynamical systems modeling: An application to the regulation of intimacy and disclosure in marriage. In T. A. Walls & J. L. Schafer (Eds.), *Models for intensive longitudinal data* (pp. 195–218). New York: Oxford University Press.

Boker, S. M., & Laurenceau, J. P. (2007). Coupled dynamics and mutually adaptive context. In T. D. Little, J. A. Bovaird, & N. A. Card (Eds.), *Modeling contextual effects in longitudinal studies* (pp. 299–324). Mahwah, NJ: Lawrence Erlbaum.

Bowlby, J. (1969). *Attachment and loss: Volume 1. Attachment*. New York: Basic Books.

Bowlby, J. (1973). *Attachment and loss: Volume 2. Separation: Anxiety and anger*. New York: Basic Books.

Bowlby, J. (1980). *Attachment and loss: Volume 3. Loss: Sadness and depression*. New York: Basic Books.

Brewer, M.B. (2000). Research design and issues of validity. In H. T. Reis & C. M. Judd (Eds.), *Handbook of research methods in social psychology* (pp. 3–16). New York: Cambridge University Press.

Campbell, L. (2005). Responses to verifying and enhancing appraisals from romantic partners: The role of trait importance and trait visibility. *European Journal of Social Psychology, 35*, 663–675.

Campbell, L., & Kashy, D. A. (2002). Estimating actor, partner, and interaction effects for dyadic data using PROC MIXED and HLM: A user-friendly guide. *Personal Relationships, 9*, 327–342.

Campbell, L., Lackenbauer, S. D., & Muise, A. (2006). When is being known or adored by romantic partners most beneficial? Self-perceptions, relationship length, and responses to partner's verifying and enhancing appraisals. *Personality and Social Psychology Bulletin, 32*, 1283–1294.

Fitzsimons, G. M., & Bargh, J. A. (2003). Thinking of you: Nonconscious pursuit of interpersonal goals associated with relationship partners. *Journal of Personality and Social Psychology, 84*, 148–164.

Fraley, R. C., & Shaver, P. R. (1998). Airport separations: A naturalistic study of adult attachment dynamics in separating couples. *Journal of Personality and Social Psychology, 75*, 1198–1212.

Gonzalez, R., & Griffin, D. (1997). On the statistics of interdependence: Treating dyadic data with respect. In S. Duck (Ed.), *Handbook of personal relationships: Theory, research and interventions* (pp. 271–302). Hoboken, NJ: John Wiley & Sons.

Gonzalez, R., & Griffin, D. (1999). The correlational analysis of dyad-level data in the distinguishable case. *Personal Relationships, 6*, 449–469.

Gonzalez, R., & Griffin, D. (2002). Modeling the personality of dyads and groups. *Journal of Personality, 70*, 901–924.

Gottman, J. M., Murray, J. D., Swanson, C. C., Tyson, R., & Swanson, K. R. (2002). *The mathematics of marriage: Dynamic nonlinear models*. Cambridge, MA: MIT Press.

Griffin, D., & Gonzalez, R. (1995). Correlational analysis of dyad-level data in the exchangeable case. *Psychological Bulletin, 118*, 430–439.

Hazan, C., & Shaver, P. (1987). Romantic love conceptualized as an attachment process. *Journal of Personality and Social Psychology, 52*, 511–524.

Kashy, D. A, Campbell, L., & Harris, D. W. (2006). Advances in data analytic approaches for relationships research: The broad utility of hierarchical linear modeling. In A. L. Vangelisti & D. Perlman (Eds.), *The Cambridge handbook of personal relationships* (pp. 73–89). New York: Cambridge University Press.

Kashy, D. A., & Kenny, D. A. (2000). The analysis of data from dyads and groups. In H. T. Reis & C. M. Judd (Eds.), *Handbook of research methods in social psychology* (pp. 451–477). New York: Cambridge University Press.

Kashy, D. A., & Snyder, D. K. (1995). Measurement and data analytic issues in couples research. *Psychological Assessment, 7*, 338–348.

Kelley, H. H., Berscheid, E., Christensen, A., Harvey, J. H., Huston, T. L., Levinger, G., et al. (1983). *Close relationships*. New York: Freeman.

Kelley, H. H., & Thibaut, J. W. (1978). *Interpersonal relations: A theory of interdependence*. New York: Wiley.

Kenny, D. A. (1994). Using the social relations model to understand relationships. In R. Erber & R. Gilmour (Eds.), *Theoretical frameworks for personal relationships* (pp. 111–127). Hillsdale, NJ: Lawrence Erlbaum.

Kenny, D. A. (1995). The effect of nonindependence on significance testing in dyadic research. *Personal Relationships, 2*, 67–75.

Kenny, D. A. (1996). Models of non-independence in dyadic research. *Journal of Social and Personal Relationships, 13*, 279–294.

Kenny, D. A., & Cook, W. (1999). Partner effects in relationship research: Conceptual issues, analytical difficulties, and illustrations. *Personal Relationships, 6*, 433–448.

Kenny, D. A., Kashy, D. A., & Bolger, N. (1998). Data analysis in social psychology. In D. T. Gilbert, S. T. Fiske, & G. Lindzey (Eds.), *The handbook of social psychology* (pp. 233–265). New York: McGraw-Hill.

Kenny, D. A., Kashy, D. A., & Cook, W. L. (2006). *Dyadic data analysis*. New York: Guilford Press.

Kenny, D. A., & La Voie, L. (1984). The social relations model. In L. Berkowitz (Ed.), *Advances in experimental social psychology* (Vol. 18, pp. 141–182). Orlando, FL: Academic Press.

Kurdek, L. A. (2003). Methodological issues in growth-curve analyses with married couples. *Personal Relationships, 10*, 235–266.

Lackenbauer, S. D., Campbell, L., Rubin, H., Fletcher, G. J. O., & Troister, T. (2010). Seeing clearly through rose-colored glasses: The unique and combined benefits of being perceived accurately and in a positively biased manner by romantic partners. *Personal Relationships, 17*, 475–492.

Levinger, G. (1980). Toward the analysis of close relationships. *Journal of Experimental Social Psychology, 16*, 510–544.

McGrath, J. E. (1982). Dilemmatics: The study of research choices and dilemmas. In J. E. McGrath, J. Martin, & R. A. Kulka (Eds.), *Judgment calls in research* (pp. 69–102). London: Sage.

Messick, D. M., & Cook, K. S. (1983). *Equity theory: Psychological and sociological perspectives*. New York: Praeger.

Murray, S. L., Rose, P., Bellavia, G. M., Holmes, J. G., & Kusche, A. G. (2002). When rejection stings: How self-esteem constrains relationship-enhancement processes. *Journal of Personality and Social Psychology, 83*, 556–573.

Neff, L. A., & Karney, B. R. (2005). To know you is to love you: The implications of global adoration and specific accuracy in marital relationships. *Journal of Personality and Social Psychology, 88*, 480–497.

Reis, H. T. (2007). Steps toward the ripening of relationship science. *Personal Relationships, 14*, 1–23.

Reis, H. T., Clark, M. S., & Holmes, J. G. (2004). Perceived partner responsiveness as an organizing construct in the study of intimacy and closeness. In D. J. Mashek & A. Aron (Eds.), *Handbook of closeness and intimacy* (pp. 201–225). Mahwah, NJ: Lawrence Erlbaum.

Reis, H. T., & Shaver, P. (1988). Intimacy as an interpersonal process. In S. W. Duck (Ed.), *Handbook of personal relationships* (pp. 367–389). Chichester, England: Wiley.

Rempel, J. K., Holmes, J. G., & Zanna, M. P. (1985). Trust in close relationships. *Journal of Personality and Social Psychology, 49*, 95–112.

Rholes, W. S., Simpson, J. A., Campbell, L., & Grich, J. (2001). Adult attachment and the transition to parenthood. *Journal of Personality and Social Psychology, 81*, 421–435.

Runkel, P. J., & McGrath, J. E. (1972). *Research on human behavior: A systematic guide*. New York: Holt, Rinehart & Winston.

Rusbult, C. E. (1980). Commitment and satisfaction in romantic associations: A test of the investment model. *Journal of Experimental Social Psychology, 16*, 172–186.

Rusbult, C. E. (1983). A longitudinal test of the investment model: The development (and deterioration) of satisfaction and commitment in heterosexual involvements. *Journal of Personality and Social Psychology, 45*, 101–117.

Simpson, J. A., Rholes, W. S., & Phillips, D. (1996). Conflict in close relationships: An attachment perspective. *Journal of Personality and Social Psychology, 71*, 899–914.

Thibaut, J. W., & Kelley, H. H. (1959). *The social psychology of groups*. New York: Wiley.

Walster, E., Walster, G. W., & Berscheid, E. (1978). *Equity: Theory and research*. Boston: Allyn & Bacon.

West, T. V., Popp, D., & Kenny, D. A. (2008). A guide for the estimation of gender and sexual orientation effects in dyadic data: An actor-partner interdependence model approach. *Personality and Social Psychology Bulletin, 34*, 321–336.

CHAPTER 2

Attachment Anxiety and Motivational Patterns in Close Relationships

PHILLIP R. SHAVER

University of California-Davis

MARIO MIKULINCER

Interdisciplinary Center (IDC) Herzliya

Ever since 1987, when Hazan and Shaver first suggested that adolescent and adult romantic love—often the first step toward what evolutionary psychologists call pair-bonding—could be illuminated by Bowlby and Ainsworth's attachment theory (Ainsworth, Blehar, Waters, & Wall, 1978; Bowlby, 1982), research on attachment processes in couple relationships and individual differences in attachment style has mushroomed. We (Mikulincer & Shaver, 2007a) reviewed much of this research in *Attachment in Adulthood: Structure, Dynamics, and Change.* A few years earlier, Edelstein and Shaver (2004) summarized evidence concerning one of the two major kinds of attachment insecurity, avoidant attachment, which seemed to embody a contradiction. It seemed odd for "attachment," which implies closeness, to be "avoidant," which implies distance. The contradiction was explained by the fact that avoidant people, like all human beings, need social relationships but feel uncomfortable being dependent on or intimately self-disclosing to anyone, even a chosen mate. Avoidant people are, in short, attached to their relationship partners without being fully open to or consciously dependent on them.

There has not been a similar summary and analysis of research on anxious attachment, the other major form of attachment insecurity, which embodies intriguing complexities and contradictions of its own. Here, we analyze and integrate research on anxious attachment, focusing on three of its characteristic motivational patterns: (1) wanting too much, too soon, (2) wanting but fearing closeness and interdependence, and (3) wanting security but feeling unable to attain or sustain it. We begin with a brief overview of adult attachment theory and research, which provides a backdrop for our analysis of anxious attachment. We then consider each of the three motivational patterns in turn. At the end, we place these patterns in a broader nature–nurture framework and briefly enumerate some unsolved problems for future conceptual and empirical research.

Overview of Adult Attachment Theory

Attachment theory (Bowlby, 1982) was originally created to explain why separation from or loss of key nurturing figures early in childhood so often has adverse consequences later in development. Bowlby viewed delinquency, failed close relationships, and individual psychopathology as outcomes of dysfunctional, hurtful early relationships with caregivers. Ainsworth and her colleagues (1978) provided a strong research base for Bowlby's theory and discovered that, even beyond the negative effects of separation and abandonment, nonoptimal parenting tilts a child's development in the direction of attachment anxiety or avoidance. Many subsequent studies, including 20-year longitudinal ones, have supported and elaborated upon Ainsworth et al.'s early insights. This research is extensively reviewed in the second edition of the *Handbook of Attachment: Theory, Research, and Clinical Applications* (Cassidy & Shaver, 2008).

For present purposes, the theory can be briefly outlined in terms of a few basic concepts. First, Bowlby (1982) assumed that attachment behaviors—being near and often in direct physical contact with one or a few key people ("attachment figures") when one is frightened, signaling to an attachment figure that emotional support is desired (e.g., by crying, calling, reaching to be picked up, hugged, or have one's hand held) when one feels threatened, and managing to reduce or control negative emotions when in the presence of an attachment figure—are the result of an innate "attachment behavioral system." Second, experiences with attachment figures are remembered and represented in "internal working models"—cognitive/affective structures that allow a person to anticipate, often unconsciously, how another person will react to one's expressions of need. According to Bowlby (1973), these internal working models are the heart of attachment patterns that can be measured in adolescents and adults through interviews (Hesse, 2008), self-report questionnaires (e.g., Brennan, Clark,

& Shaver, 1998), and behavioral observations (e.g., Simpson, Rholes, & Nelligan, 1992). In social and personality psychology, these patterns are often called attachment "styles" or "orientations."

The initial methods of measuring attachment styles in adolescence and adulthood were typological (e.g., Bartholomew & Horowitz, 1991; Hazan & Shaver, 1987), because Ainsworth et al. (1978) classified infants into three categories—secure, anxious, and avoidant. However, subsequent studies (e.g., Brennan et al., 1998; Fraley & Waller, 1998) revealed that attachment orientations are best conceptualized as regions in a two-dimensional space, with no sharp category boundaries or distinct types. The first dimension, *attachment-related anxiety*, represents the degree to which a person worries that relationship partners will be unavailable in times of need, become interested in someone else, or end the relationship. The second dimension, *attachment-related avoidance*, reflects the extent to which a person distrusts relationship partners' goodwill and strives to maintain independence and emotional distance from partners.

The two attachment-style dimensions can be measured with the 36-item Experiences in Close Relationships inventory (ECR; Brennan et al., 1998), which is reliable in both the internal-consistency and test–retest senses and has high construct, predictive, and discriminant validity (Mikulincer & Shaver, 2007a). A total of 18 items assess the anxiety dimension (e.g., "I need a lot of reassurance that I am loved by my partner"; "I resent it when my partner spends time away from me"), and 18 assess the avoidance dimension (e.g., "I try to avoid getting too close to my partner"; "I prefer not to show a partner how I feel deep down"). The two scales were conceptualized as independent and have generally been found to be empirically uncorrelated.

Hundreds of studies using the ECR and other self-report measures of adult attachment have found that higher scores on attachment anxiety or avoidance are associated in theoretically predictable ways with lower self-esteem, more negative views of others, poorer relationship quality and mental health, and problems in interpersonal interactions, coping with stress, and adjustment (see Mikulincer & Shaver, 2007a, for a review). In the present chapter, as we explained at the outset, we focus mainly on attachment anxiety and its implications for a person's motivational patterns in close relationships.

Attachment Anxiety and Hyperactivated Behavioral Strategies

Attachment theory predicts that when individuals encounter psychological or social threats they will seek either actual or symbolic support from an attachment figure (Mikulincer & Shaver, 2003, 2007a). "Actual" support seeking involves signaling to and moving toward a familiar attachment figure. "Symbolic" support seeking includes calling upon memories of interactions with actual attachment figures or relying, in one's mind, on imagined attachment figures, such as imaginary friends or religious figures (e.g., God or gods, guardian angels, or deceased ancestors). If these strategies achieve the desired level of comfort and security, the need is satisfied and the person is able to turn attention away from support seeking and toward renewed engagement with the world. If they do not achieve the desired level of comfort and security, however, people may either "hyperactivate" or "deactivate" their attachment behavioral system. Deactivating strategies, such as denying threats, denying one's wish or need for help, and thinking only of one's capacity to handle threats alone, are characteristic of people who score high on measures of attachment-related avoidance (Cassidy & Kobak, 1988; Mikulincer & Shaver, 2003). In contrast, hyperactivating strategies, such as ruminating on threats, energetically seeking or demanding support, and expressing intense emotional distress, are characteristic of people who score high on measures of attachment-related anxiety (Mikulincer & Shaver, 2003).

Following Bowlby (1982), we (Mikulincer & Shaver, 2007a) conceptualized hyperactivating strategies mainly as "fight" (persist or protest) responses that intensify the attachment behavioral system's primary strategy—proximity seeking—to the point where it can create relational problems. (For example, it may drive a relationship partner away rather than eliciting sympathetic support.) The main goal of these strategies is to compel an attachment figure, viewed as unreliable or insufficiently available and responsive, to pay more attention and provide better protection or support. Anxious individuals attempt to attain this goal by maintaining their attachment system in a chronically activated state until an attachment figure is perceived to be adequately available and responsive. This hyperactivation includes attempts to elicit a partner's attention, involvement, and support through clinging and controlling responses (e.g., begging for or demanding more attention, nervously scanning the partner's computer files to check for signs of other sexual or love interests, insisting on going with the partner to places and events to which the partner had planned to go alone) as well as engaging in cognitive and behavioral efforts to minimize psychological distance from the partner (Shaver & Mikulincer, 2002).

These efforts to increase closeness can be aimed not only at establishing actual physical contact but also at achieving a psychological sense of self–other similarity, intimacy, or "oneness" (Mikulincer & Shaver, 2003). Hyperactivating strategies can also include being overly dependent on relationship partners and viewing oneself as relatively helpless and incompetent without their help (Mikulincer & Florian, 1998).

Hyperactivating strategies cause a person to maintain vigilance toward possible threats and signs of an attachment figure's unavailability, the two primary "triggers" that activate the attachment system (Bowlby, 1973). Once these issues become the focus of individuals' attention, they more or less guarantee that the attachment system will remain continuously active. Because anxious hyperactivation of the attachment system intensifies rumination on threats and amplifies emotional reactions to threats, attachment-anxious people's psychological pain is exacerbated rather than subdued, and their doubts about being able to achieve relief are strengthened. Moreover, because signs of rejection and abandonment are viewed as threats in their own right, hyperactivating strategies foster anxious, hypervigilant attention to relationship partners and rapid detection of possible signs of disapproval, waning interest, or impending abandonment. In this way, hyperactivating strategies produce a self-amplifying cycle of distress in which chronic attachment-system activation interferes with engagement in non-attachment-related activities, such as exploration and learning (Mikulincer, 1995).

Based on the work of Ainsworth et al. (1978), we (Mikulincer & Shaver, 2007a) suggested that hyperactivated attachment behavior is learned in a particular kind of social environment during infancy and early childhood, where the primary attachment figure is sometimes responsive and at other times not. This kind of environment places a person on a partial reinforcement schedule for being vigilant, for noisily demanding attention, for exaggerating needs and vulnerabilities, and for demanding better care. Unfortunately, although this strategy is sometimes successful in gaining the attention and help of the primary attachment figure (e.g., a parent) who is also anxious, self-preoccupied, and nonresponsive, it can erode later relationship partners' patience and affection, resulting in just the kind of rejection and abandonment against which the anxious person was defending. What may have worked with an early attachment figure does not necessarily work with a new relationship partner.

Research has demonstrated that attachment anxiety is associated with exaggeration of threats, negative self-views, and pessimistic beliefs and doubts about other people (e.g., Bartholomew & Horowitz, 1991; Mikulincer, 1995; Mikulincer & Florian, 1998). Individuals who score high on measures of attachment anxiety have ready access to painful memories and are victims of an automatic spread of negative affect from one remembered incident to another (e.g., Mikulincer & Orbach, 1995). Moreover, their attachment-related worries are often aroused even when there is no external threat (e.g., Mikulincer, Gillath, & Shaver, 2002).

There is also evidence that attachment anxiety biases the way people perceive their relationship partners (see Mikulincer & Shaver, 2007a, for a review). Although people who score high on attachment anxiety have a history of frustrating interactions with attachment figures, they nevertheless believe that if they intensify their proximity-seeking efforts, they may cause relationship partners to pay more attention and provide more adequate support (Cassidy & Berlin, 1994). Hence, they do not form simple or uniformly negative impressions of others, because that would entail concluding that proximity seeking is hopeless. Rather, even when angry they are likely to accept some of the blame for their partner's unreliable attention and care (Mikulincer & Shaver, 2003). This can lead to contradictory appraisals of a partner's potential value and sometimes results in failure to opt out of a violent relationship.

Attachment research indicates that attachment-anxious people's wishes and goals reflect both their intense need for closeness and their fear of separation and abandonment. For example, two researchers (Avihou, 2006; Raz, 2002) used the Core Conflictual Relationship Theme method (Luborsky & Crits-Christoph, 1998) to code wishes revealed in study participants' narratives of interpersonal interactions or in their dreams, finding that self-reports of attachment anxiety were associated with wishes to be loved, validated, and accepted by others. In a study of correspondences between attachment theory and Kohut's (1977) self psychology, Banai, Mikulincer, and Shaver (2005) found that self-reports of attachment anxiety were associated with relatively high scores on measures of Kohut's concepts of hunger for mirroring (wishing to be admired for one's qualities and accomplishments), idealization (forming idealized images of relationship partners and then wishing to merge with them), and twinship (feeling similar to others and being accepted by them in close relationships).

Beyond having an unusually strong desire for closeness, attachment-anxious individuals experience higher than usual levels of rejection sensitivity—the tendency to anticipate and overreact to rejection (e.g., Downey & Feldman, 1996)—and are quicker to recognize rejection-related words in lexical decision tasks (Baldwin & Kay, 2003; Baldwin & Meunier, 1999). Moreover, attachment-anxious people have difficulty inhibiting rejection-related thoughts (Baldwin & Kay, 2003; Baldwin & Meunier, 1999). Baldwin and Kay (2003), for example, exposed study participants to tones paired with photographs of rejecting (frowning) or accepting (smiling) faces and then administered a lexical decision task in which rejection-related words were paired with each of the tones. Secure people were slower to react to rejection-related words even when they were paired with rejection tones (compared with a neutral tone), and anxious people reacted faster to these words even in the presence of the acceptance tone. This implies that anxious individuals are so hypervigilant with respect to disapproval and rejection that they entertain rejection-related thoughts even in the presence of an accepting interpersonal context.

In short, attachment anxiety is associated with important goals (insisting on proximity to an attachment figure, worrying about and seeking to avoid separation and rejection), which go hand in hand with particular cognitive, affective, and behavioral processes aimed at attaining these goals. Pursuit of these goals can be characterized in terms of three distinct motivational patterns, to which we devote the next section of this chapter.

What Do People Who Score High on Attachment Anxiety Want in Close Relationships?

Research on anxious attachment suggests that people plagued by this form of insecurity are likely to exhibit one or more of three motivational patterns: (1) wanting too much, too soon; (2) wanting but fearing closeness and interdependence; and (3) wanting security but feeling unable to attain or sustain it. We will consider each of these patterns in turn.

Wanting Too Much, Too Soon

Attachment-anxious individuals' intense hunger for love, closeness, and merger can cause them to push a partner too soon and too insistently for intimacy, a sense of sharing, togetherness, and communion, and a long-term commitment. Paradoxically and tragically, this often provokes distancing and rejection on the other person's part, which is exactly what anxious persons hope to avoid. In the course of a romantic relationship, couples usually transition gradually from falling in love to loving each other (Berscheid & Regan, 2005). Flirtation gives way to more temporally extended joint activities, partners begin to share more intimate information and more complete narratives of their personal histories and concerns, and gradually these intimate exchanges are supplemented by discussions of long-term goals and plans the couple might pursue together. In addition, the importance of support, nurturance, and intimacy as determinants of relationship quality increases as early infatuation recedes in importance. However, the consolidation of a lasting and committed relationship—a true attachment or pair-bond—takes time (Hazan & Zeifman, 1999, estimated that establishing a full-blown attachment takes more than a year) and is a gradual process that must be coordinated between the two partners. During this process, attachment-related anxiety can interfere with normal relationship development, because strong needs for closeness and an intense fear of separation or rejection cause anxious people to try to speed up the commitment process without taking into account the partner's needs, preferences, and concerns.

Evidence for anxious people's "too much, too soon" motivational pattern is found in studies examining differences in the way people disclose personal information to a partner. Although both anxious and relatively secure individuals share a willingness to self-disclose and to react with interest to a partner's self-disclosures, the anxious ones tend to disclose indiscriminately, even before their relationship partner is prepared to match a particular level of intimacy (Mikulincer & Nachshon, 1991). Moreover, they tend to be relatively unresponsive to their partner's disclosures while focusing mainly on their own feelings and disclosures (Grabill & Kerns, 2000; Mikulincer & Nachshon).

Although disclosing too much, too soon may temporarily quell fears of separation and distance and increase the sense of deep interpersonal similarities where no such similarities have been established, it may fail to engender reciprocal intimacy and can evoke negative reactions in a partner who feels misunderstood and invalidated (Reis & Shaver, 1988). This destructive sequence of events was observed by Bradford, Feeney, and Campbell (2002), who asked dating couples to rate each conversation they had with each other over a 7-day period, focusing on the amount and emotional tone of self-disclosures and their satisfaction with the disclosure process. The data indicated that anxious participants self-disclosed more, but their partners rated these disclosures as relatively unsatisfying and often negative in tone.

The too much, too soon pattern has also been noted in studies of the association between attachment anxiety and commitment. Senchak and Leonard (1992), for example, found that anxious men acquired marriage licenses much sooner (after 19 months of courtship) than secure men (49 months). Morgan and Shaver (1999), reviewing this and other early studies on the topic, concluded that anxious people tend to commit prematurely, before getting to know their partner well, which leaves them more vulnerable to either exploitative partners or ones who feel compelled to back out of what they see as likely to become a suffocating relationship. Indeed, in the very first study of "romantic attachment," Hazan and Shaver (1987) found that anxious individuals were more likely than their peers to agree with the statement, "Few people are as willing and able as I am to commit themselves to a long-term relationship." In other words, based on their experience, which they undoubtedly helped to shape, they thought other people were too reluctant to commit.

A similar pattern is evident in the ways anxious people negotiate closeness and distance. According to Pistole (1994), their unmet needs for love and merger cause them to seek closeness to such an extent that it makes their partners uncomfortable. Moreover, their strong fear of rejection may cause them to misinterpret a partner's desire for privacy or autonomy as a sign of rejection, which tempts them to escalate demands for intimacy to such an extent that their partner withdraws or flees. This unwelcome series of events results in what marital therapists call a "demand-withdrawal" (Christensen, 1988) or "pursuit-withdrawal" (Bartholomew & Allison, 2006) cycle—one of the major predictors of relationship distress and dissolution.

Attachment-anxious people's difficulty in regulating closeness and distance within their romantic relationships was documented in a recent study by Lavy, Mikulincer, and Shaver (2010). We asked young adults to complete self-report scales measuring attachment anxiety, intrusive behavior in their couple relationships (e.g., asking their partners very personal questions and interfering with their plans to engage in some activities alone), and subjective experiences of being intrusive. As expected, more anxious people reported greater engagement in intrusive behavior, and they experienced more negative feelings when being intrusive, probably because they sensed that their intrusive behavior was motivated by fear of rejection and a lack of partner affection. However, attachment anxiety was also associated with feeling appreciated by a partner, perhaps reflecting anxious individuals' pleasure and gratitude when intrusive responses sometimes increased a partner's attention and apparent concern.

In another study, Lavy (2006) asked members of seriously dating couples to complete a measure of attachment anxiety and daily measures of relationship satisfaction and intrusive behavior over a 14-day period. More anxious participants typically acted more intrusively, but this association was moderated by relationship satisfaction on the previous day. That is, anxious people reported relatively high levels of intrusive behavior mainly when they were dissatisfied with their relationship on the previous day. These findings imply that daily intrusive behaviors may be part of anxious people's efforts to create or restore a desired level of closeness. Like other examples of the too much, too soon motivational pattern, these closeness-regulation efforts are likely to backfire if they cause a partner to feel suffocated or intruded upon.

Shaver, Schachner, and Mikulincer (2005) observed a similar pattern of daily behavior in their study of excessive reassurance seeking from a romantic partner. Both members of 61 dating couples independently logged onto a website each evening for a period of 14 days and answered questions about daily disagreements with their partner, daily reassurance seeking, and daily affect or mood. Participants who scored higher on attachment anxiety were more likely to seek daily reassurance from their dating partner. In addition, more anxious participants sought more reassurance following days on which they experienced relationship conflicts than following days on which little or no conflict was reported. Thus, it seems that anxious people seek reassurance as a way of quelling worries that interfere with their sense of security and being loved.

A more general characterization of the too much, too soon pattern is provided by studies that examined attachment-style differences in scores on the Inventory of Interpersonal Problems (IIP; Horowitz, Rosenberg, Baer, Ureno, & Villasenor, 1988). In this inventory, people are asked about the extent to which they feel they are overly autocratic, competitive, cold, introverted, exploitable, nurturant, and expressive/demanding. Across several studies and across both self-reports and reports from friends, attachment anxiety has been associated with higher levels of problems related to being overly expressive and demanding (e.g., Bartholomew & Horowitz, 1991; Bookwala & Zdaniuk, 1998).

Some studies of this kind (e.g., Bookwala & Zdaniuk, 1998; Chen & Mallinckrodt, 2002) indicate that more anxious individuals are more likely to report being both overly expressive and overly nonassertive, which again seems to imply a contradiction or a major inconsistency. These findings suggest a second motivational pattern worth exploring: wanting and needing a great deal, and sometimes expressing this intense need directly, but also being (perhaps justifiably) worried that they will be disapproved or rejected if they assert their needs too forcefully. This conflict is the topic of the next section.

The "Want but Fear" Pattern

Strong needs for love and closeness, combined with intense fears of rejection, can create a want but fear pattern, marked by conflicting relational goals and action tendencies and ambivalent attitudes toward relationship partners. Attachment-anxious people are strongly influenced by their desires for security and closeness, which causes them to focus, at times, on the potential rewards of intimacy and to entertain positive attitudes toward relationships and relationship partners. They are also affected, however, by fears of rejection and abandonment and by memories of frustrating attachment relationships, which lead them to exaggerate the risks of closeness (e.g., Boon & Griffin, 1996) and the potentially negative traits and intentions of their partners (e.g., Mikulincer & Horesh, 1999). Ainsworth et al. (1978) originally called the anxious attachment pattern in infants "insecure/ambivalent," and Hazan and Shaver (1987) initially labeled the equivalent adult pattern "anxious/ambivalent." Over time, mainly for simplicity but also in deference to the dimensional rather than the typological approach to measurement, anxious ambivalence gave way to the simpler "attachment anxiety."

In one of the first studies of the want but fear motivational pattern, Davis, Shaver, and Vernon (2004) found that adults scoring high on attachment anxiety wanted both more sex and less sex under certain conditions—more when they thought about sex as a barometer of their partner's love and commitment, less when they thought about the unpleasantness of acceding to a partner's sexual demands in an effort to avoid disapproval or rejection. Additional evidence was provided by Bartz and Lydon (2006), who assessed patterns of interaction with an attractive, opposite-sex confederate in a laboratory setting. They found that attachment anxiety was associated with a stronger desire to increase closeness to the new partner, but when the confederate explicitly expressed interest in greater closeness, more anxious participants felt nervous and performed worse on a cognitive task. In contrast, less anxious people accepted the confederate's favorable comments with fewer worries and performed better on the task. Thus, although anxious individuals are interested in developing relationships with attractive people, when such people show signs of interest, it may arouse fears of being rejected or disliked.

More direct evidence for the want but fear pattern is provided by a recent series of studies (Mikulincer, Shaver, Bar-On, & Ein-Dor, 2010) on attachment-style differences in relational ambivalence—the simultaneous activation of conflicting motivational forces (want/approach and fear/avoid) and attitudes (positive and negative) when people think about close relationships. In the first study, participants completed a self-report measure of attachment anxiety as well as measures of explicit and implicit relational ambivalence. Explicit relational ambivalence was assessed with the Ambivalence in Intimate Relationships scale (AIR; Thompson & Holmes, 1996), which asks people to describe their positive and negative feelings toward several traits of a romantic partner. To assess implicit relational ambivalence, we presented participants with a set of positive and negative words related to closeness to and distance from relationship partners (e.g., closeness, intrusiveness, privacy, loneliness). On each trial, participants were asked either to pull a lever toward themselves (an approach response) or push the lever away from themselves (an avoidance response) when they encoded the meaning of a word. The time taken to initiate each response was automatically recorded. Ambivalence toward a word (or the concept named by the word) was assessed in terms of rapid approach and avoidance responses to relationship words and in terms of smaller differences between such latencies. As expected, attachment anxiety was associated with both explicit and implicit measures of relational ambivalence.

In a second study, we asked participants to complete a new scale designed to measure explicit motivational ambivalence toward closeness and distance goals in a romantic relationship and to perform the approach–avoidance (pull–push) task while focusing attention on a particular romantic partner. At both the explicit and implicit levels, attachment anxiety was associated with strong approach–avoidance ambivalence toward *closeness* to a romantic partner. However, no significant association was found between attachment anxiety and either explicit or implicit ambivalence toward relational *distance*.

In a subsequent study (Mikulincer et al., 2010, Study 4), we examined the possibility that attachment-anxious people's want but fear motivational pattern would be intensified by relational contexts that activated just one kind of relational tendency (either approach or avoidance). If attachment anxiety is associated with simultaneous activation of antagonistic intimacy-related goals and attitudes toward a relationship partner, these goals and attitudes should be connected such that activation of one goal or attitude paradoxically causes activation of the antagonistic or opposite goals and attitudes. That is, attachment-anxious people should react to a partner's expression of interest and love not only with approach tendencies and a positive view of the loving partner but also with fears of rejection and abandonment and doubts about the partner's intentions. Moreover, they should react to a partner's signals of disinterest and rejection not only with avoidance tendencies but also with intense wishes for affection and security and hopes that the partner will be responsive.

To test these ideas, we asked participants to perform a guided imagination task after being randomly assigned to one of three experimental conditions: (1) relationship initiation (imagining the beginning of a romantic relationship), (2) separation (imagining the breakup of a romantic relationship), or (3) a control condition (imagining watching a TV show). Participants then completed the approach–avoidance (pull–push) task as a measure of implicit relational ambivalence. We found that the association between attachment anxiety and implicit ambivalence toward closeness-related words was stronger in the relationship initiation and separation conditions than in the control condition. However, no significant association was found between attachment anxiety and implicit ambivalence toward distance words in any of the three experimental conditions.

Next, we (Mikulincer et al., 2010, Studies 5–6) examined the simultaneous activation of approach (want) and avoidance (fear) motives. That is, each study consisted of the experimental activation of one of the two kinds of motives (approaching relational closeness/distance or avoiding relational closeness/distance) through guided imagination exercises, followed by an assessment of the implicit activation of the opposing motive. We found that the greater individuals' attachment anxiety, (1) the more their implicit motive to avoid closeness was aroused by considering the positive aspects of relational closeness, and (2) the more their implicit motive to approach closeness was aroused by considering the negative aspects of relational closeness.

These findings revealed two relationship contexts that triggered attachment-anxious people's want but fear pattern. First, this pattern seemed to be particularly activated when people thought about separation from a romantic partner and about the negative aspects of close relationships. In those situations, which typically fostered fears of separation and abandonment, attachment-anxious people exhibited a paradoxical implicit tendency to approach relational closeness. For them, experimental manipulations that made negative aspects of a relationship salient also seemed to activate preconscious representations of its positive aspects.

This kind of response is consistent with Aron, Aron, and Allen's (1998) finding that unrequited or unreciprocated love is more common among people with an anxious attachment style. It is also consistent with findings that attachment-anxious people tend to break up and then get back together again with the same partner multiple times (Kirkpatrick & Hazan, 1994) and are more likely than their less anxious counterparts to feel sexually attracted to their former partner and more likely to become sexually reinvolved (Davis, Shaver, & Vernon, 2003). Similarly, Henderson, Bartholomew, and Dutton (1997) found that abused women who scored higher on attachment anxiety were more likely to remain sexually and emotionally involved with their abusive partner following separations.

The second relationship context that activates the want but fear pattern is a situation that emphasizes positive and gratifying aspects of close relationships. Under these conditions, attachment-anxious participants in our studies exhibited a paradoxical preconscious tendency to avoid relational closeness. This observation resembles Mikulincer and Sheffi's (2000) findings concerning the effects of a positive mood induction on creative problem solving. In their study, attachment-anxious participants reacted to positive experiences (retrieving a happy memory or watching a brief comedy film) with *impaired* creative problem solving, which was the opposite of what occurred with secure individuals and the same as what had happened in previous studies following negative mood inductions. It seems that attachment-anxious people are prone to turn a positive signal into a harbinger of danger. Perhaps they initially experience positive feelings but then are quickly reminded of the downside of previous relationship experiences that began positively but ended painfully. Once attuned to these negative memories, anxious people may suffer from a spread of negative associations that interfere with creativity (in Mikulincer and Sheffi's studies) and heightens the want but fear pattern (in Mikulincer et al.'s 2010 studies).

It is worth emphasizing that attachment anxiety was more strongly and consistently associated with motivational ambivalence toward closeness than toward distance. That is, attachment-anxious individuals' ambivalence in romantic relationships is more focused on desires for proximity and support and fears of rejection and separation than on desires for solitude and independence and fears of loneliness and isolation. This is not surprising, because attachment-anxious people tend to overemphasize their dependence on a partner's availability and responsiveness (Shaver & Mikulincer, 2002) and worry about gaining love and support and avoiding rejection and separation. They are less interested in the potentially positive aspects of autonomy and solitude. Due to their sense of vulnerability and helplessness, attachment-anxious people may experience distance from others as dangerous and therefore tend to emphasize the negative aspects of being alone (Mikulincer & Shaver, 2007a).

The "Want but Cannot" Pattern

Anxious people's fears of rejection and doubts about their own value and competence can sometimes inhibit the translation of their relational wishes and goals into effective intentions, plans, and actions. Due to their fears and doubts, they sometimes defer expression of their needs so as not to experience the pain of disapproval or rejection. This is another way they are plagued by conflicts, because at times they are overly insistent and at other times less assertive, with neither strategy leading consistently to favorable outcomes. In this way, their want but cannot motivational pattern engenders hopelessness, low self-esteem, and pessimism.

The dissociation between motivation and behavior can be seen in different kinds of studies of attachment-style differences in support seeking. When *preconscious* support-seeking ideation is primed (e.g., Mikulincer, Gillath, & Shaver, 2002), anxious individuals, like secure ones, react to subliminal threats (words such as death, illness, and separation) with heightened availability of attachment figures' names. Moreover, they exhibit such activation even in the absence of threatening primes, implying that attachment-anxious people are preconsciously motivated to seek support from attachment figures even in nonthreatening situations.

However, studies assessing *conscious* self-reports of support-seeking tendencies yield different results. Several studies have found attachment anxiety to be associated with *lower* levels of support seeking (e.g., Mikulincer & Florian, 1998; Ognibene & Collins, 1998), and others have found no association at all between anxiety and support seeking (e.g., Horppu & Ikonen-Varila, 2001; Pierce & Lydon, 1998). The inconsistent results across studies, including those summarized in the previous paragraph, suggest that anxious people's intense wish for support is countered by doubts about the availability of support or their ability to enlist others' support. Indeed, Vogel and Wei (2005) found two opposing causal pathways by which attachment anxiety was associated with support seeking. In one path, attachment anxiety was associated with greater psychological distress, which in turn heightened support seeking. In the other path, attachment anxiety was linked with negative perceptions of others' supportiveness, which led to reduced support seeking. This inhibiting effect of negative perceptions of support availability was also observed by Rholes, Simpson, Campbell, and Grich (2001). In their study, attachment-anxious pregnant women who perceived their husbands to be unsupportive 6 weeks before giving birth sought less support from their husbands 6 months postpartum. However, when anxious women perceived their husbands to be supportive, their wish for care and protection resulted in a significant increase in support seeking after delivery (compared with less anxious women). Moreover, the group of anxiously attached new mothers who felt inadequately supported by their husbands was more vulnerable to postpartum depression.

Observational studies of support seeking from dating and marital partners (e.g., Collins & Feeney, 2000; Fraley & Shaver, 1998) indicate that although attachment anxiety is not associated with direct requests for partner support, it is associated with indirect methods of support seeking, such as conveying a need for help through nonverbal distress signals (crying, pouting, or sulking). Together, the findings of these and the previously described studies show that attachment-anxious individuals are subject to preconscious activation of support-seeking tendencies, but their fear of rejection and doubts about their love-worthiness sometimes interfere with efforts to seek support. What does show up in behavior may seem to a partner to be immature nonverbal signals without meaningful content.

Sexual behavior is another arena in which the want but cannot motivational pattern plays itself out. Because sex is an important route to closeness and intimacy, attachment-anxious people generally have a positive attitude toward sex and use it as a means of fulfilling unmet needs for security and love. However, their negative working models of self and their worries about rejection and disapproval may make it difficult for them to relax and "let go" sexually, which may reduce pleasure or infuse sexuality with anxiety or distress.

Attachment anxiety is associated with less positive and more negative feelings during sex (e.g., Birnbaum, Reis, Mikulincer, Gillath, & Orpaz, 2006; Tracy, Shaver, Albino, & Cooper, 2003); lower levels of sexual arousal, intimacy, and pleasure (Hazan & Zeifman, 1994); and less positive appraisals of one's own sexual appeal (Cyranowski & Andersen, 1998). However, although attachment-anxious individuals report unsatisfying sexual interactions, they also express a strong desire for their partner's emotional involvement during sex (Birnbaum et al., 2006). With regard to sexual coercion, Tracy et al. (2003) found that higher attachment anxiety was associated with higher rates of physical coercion on the part of sexual partners and more involvement in unwanted but consensual sex, mostly on the part of women. Impett and Peplau (2002) also found that women scoring high on attachment anxiety were more accepting of unwanted sex to reduce relational conflicts and avoid rejection. Brassard, Shaver, and Lussier (2007) obtained a different pattern for anxiously attached men; they perceived their partner as avoiding sex and tried to pressure them into engaging in it more frequently. This may be a more "masculine" way of feeling slighted by one's partner and wishing for greater intimacy and care.

To provide a direct test of our hypothesis concerning attachment anxiety and the want but cannot motivational pattern, we asked 83 Israeli undergraduates (54 women, 29 men) to complete the Experiences in Close Relationships scale (ECR; Brennan et al., 1998) together with scales assessing what people want in their lives and the extent to which they believe they can fulfill their dreams. Specifically, embedded within a battery of eight questionnaires were two that dealt with wishes concerning closeness goals, independence goals, and nonrelational goals. Participants were asked to report how much they wanted to achieve these goals and how likely they thought they would be to do so. The two questionnaires were separated by three other scales to reduce the likelihood that participants would directly compare their answers to the two sets of questions.

In the want questionnaire, participants received a list of 30 wishes: 10 closeness-related wishes (e.g., to be loved, to be accepted by others), 10 independence-related wishes (e.g., to be independent, to assert oneself), and 10 nonrelational wishes (e.g., to succeed in college, to succeed at work). They were asked to rate the extent to which they wanted to realize each of these wishes in their lives. In the can questionnaire, they received the same list of 30 wishes and were asked to rate the extent to which they felt capable of realizing each of them. All of the ratings were made on a seven-point scale ranging from 1 (*not at all*) to 7 (*very much*). Half of the participants first answered the want questionnaire, and the other half first answered the can questionnaire. To overcome problems associated with an exclusively self-report methodology, we asked the best friend of 50 of the participants to complete the same scales based on their knowledge of the participants. Cronbach's alphas were high (> .82) for each category of wishes in each of the questionnaires, so we were able to compute six scores for each participant's reports and six for each of the friend's reports. We also computed a want–can discrepancy score for each category of wishes by subtracting the can score from the want score for that category.

Pearson correlation coefficients revealed similar patterns of association among the self-report scores and the friend-report scores. First, attachment anxiety was significantly associated with want scores for closeness-related wishes, $r(81) = .47$, $p < .01$ for self-reports, $r(48) = .53$, $p < .01$ for friend reports, but not with want scores for autonomy and nonrelational wishes, all r's < .14. Second, significant associations were found between attachment anxiety and lower can scores in all three categories of wishes across both self- and friend reports, r's > −.42, all p's < .01. In line with previous findings regarding appraisals of self-efficacy and self-competence (see Mikulincer & Shaver, 2007a, for a review), the higher the participants' attachment anxiety scores, the less they and their friends believed they were capable of realizing closeness, independence, and nonrelational goals. Third, significant positive associations were found between attachment anxiety and the want–can discrepancy score for closeness wishes: $r(81) = .55, p < .01$ for self-reports, $r(48) = .58, p < .01$ for friend reports. However, no significant associations were found between attachment anxiety and discrepancy scores for independence and nonrelational wishes in either self- or friend reports, r's $< .06$.

Overall, across both self-reports and friend reports, attachment anxiety was associated with larger discrepancies between wishes for closeness/intimacy and the extent to which these wishes were likely to be satisfied in real life. In other words, anxiously attached people clearly exhibited the want but cannot pattern with respect to closeness and intimacy. This want but cannot pattern was confined to closeness/intimacy wishes and did not extend to other kinds of goals. As explained in the next section, these results support previous findings suggesting that attachment anxiety is not simply redundant with neuroticism or general anxiety. Nevertheless, it shares some roots with the other kinds of anxiety.

Where Do These Motivational Patterns Come From?

It's natural to wonder whether attachment theory is correct in attributing the motivational patterns described here to experiences in relationships rather than to general personality traits such as neuroticism. In our 2007 book (Mikulincer & Shaver, 2007a), we summarized evidence on this issue, finding that although measures of attachment anxiety and neuroticism are often correlated in the range of .30 to .50, when neuroticism or other measures of general anxiety are statistically controlled, relationship-related predictions based on attachment theory continue to be supported. Nevertheless, it is worthwhile to consider the evidence for general personality trait and genetic contributions to attachment anxiety.

Brennan and Shaver (1998) were the first to show that the two dimensions underlying adult attachment styles are systematically related to the two dimensions underlying personality disorders (measured in a large college student sample using self-report measures). Moreover, the dimensions underlying the two sets of measures were related to the students' reports of childhood experiences with parents, including divorce and parenting styles, suggesting (though not proving) that there are social influences on both kinds of anxiety or insecurity. There were no measures of genetic influences in that study, however.

Crawford, Shaver, Cohen, Pilkonis, Gillath, and Kasen (2006) subsequently found again, in a 17-year longitudinal study running from adolescence (age 16) to adulthood (age 33), that self-reported attachment patterns (assessed in terms of anxiety and avoidance) were systematically related to personality disorders. In a related study, Crawford, Livesley, Jang, Shaver, Cohen, and Ganiban (2007) tested 239 twin pairs from a community sample to see how anxious and avoidant attachment were related to personality disorders (PD) and whether any of the associations were due to genetic influences. A factor analysis indicated that self-reported anxious attachment and 11 PD scales loaded on a single factor (while avoidant attachment and 4 PD scales loaded on a separate factor). Biometric models indicated that 40% of the variance in anxious attachment was heritable, and 63% of its association with the anxiety-related PD dimension was due to common genetic effects. Donellan, Burt, Levendosky, and Klump (2008) conducted a similar twin study, using the "Big Five" personality traits instead of PDs as potentially associated variables. There was a sizable correlation between anxious attachment and neuroticism, and it appeared to be due to shared genetic influences. Finally, Gillath, Shaver, Baek, and Chun (2008) conducted a candidate gene study and found that self-reported attachment anxiety was associated with a particular polymorphism of the DRD2 dopamine receptor gene, implying some genetic bases of anxious attachment.

Considering these preliminary studies, it seems likely that genetic factors influence attachment anxiety and its association with more general neuroticism and personality disorders that involve anxiety. The authors of these studies said explicitly that the fact that there are genetic influences on attachment anxiety does not mean that it is fully determined by genes rather than one's history of attachment relationships. The fact that attachment anxiety's effects occur in our experiments even when neuroticism or general anxiety is statistically controlled suggests that the shared genetic influences on attachment-related and more general anxiety do not account for all of the effects of attachment anxiety measured in the close relationship domain. Nevertheless, they raise intriguing questions about how, precisely, the motivational patterns discussed in this chapter arise in development and how the genetic and social influences combine or interact to produce them. This is an important issue for future research.

Concluding Comments

Considerable evidence has accumulated for the existence of three motivational patterns organizing the experiences and relational behavior of people who score high on measures of attachment anxiety: wanting too much intimacy and commitment too soon in a relationship, wanting closeness and a high degree of interdependence but also fearing that this may lead to a devastating loss, and wishing for secure closeness but feeling unable to attain it. Each of these patterns involves conflict and ambivalence, and each tends to produce self-defeating behavior—behavior that hastens the very rejection and abandonment feared by attachment-anxious people. The three motivational patterns are probably due partly to genetic influences that also affect neuroticism, general anxiety, and anxiety-laden personality disorders. But measures of attachment anxiety predict behavior in experiments and in actual relationships even when more general forms of anxiety are statistically controlled. The complex developmental path that interweaves anxious proclivities with social experiences remains to be mapped. Meanwhile, experimental studies aimed at enhancing a person's sense of security through priming of security-related images, memories, or cognitions (e.g., names of security-enhancing attachment figures; Mikulincer & Shaver, 2007b) fortunately indicate that attachment anxiety can be alleviated, with benefits to a person's quality of life and close relationships, which means that the genetic influences are not all powerful. Determining the precise effects of interventions on the motivational patterns described in this chapter is an important task for future research.

References

Ainsworth, M. D. S., Blehar, M. C., Waters, E., & Wall, S. (1978). *Patterns of attachment: A psychological study of the strange situation.* Hillsdale, NJ: Erlbaum.

Aron, A., Aron, E. N., & Allen, J. (1998). Motivations for unreciprocated love. *Personality and Social Psychology Bulletin, 24,* 787–796.

Avihou, N. (2006). *Attachment orientations and dreaming: An examination of the unconscious components of the attachment system.* Unpublished doctoral dissertation, Bar-Ilan University, Ramat Gan, Israel.

Baldwin, M. W., & Kay, A. C. (2003). Adult attachment and the inhibition of rejection. *Journal of Social and Clinical Psychology, 22,* 275–293.

Baldwin, M. W., & Meunier, J. (1999). The cued activation of attachment relational schemas. *Social Cognition, 17,* 209–227.

Banai, E., Mikulincer, M., & Shaver, P. R. (2005). "Selfobject" needs in Kohut's self psychology: Links with attachment, self-cohesion, affect regulation, and adjustment. *Psychoanalytic Psychology, 22,* 224–260.

Bartholomew, K., & Allison, C. J. (2006). An attachment perspective on abusive dynamics in intimate relationships. In M. Mikulincer & G. S. Goodman (Eds.), *Dynamics of romantic love* (pp. 102–127). New York: Guilford Press.

Bartholomew, K., & Horowitz, L. M. (1991). Attachment styles among young adults: A test of a four-category model. *Journal of Personality and Social Psychology, 61,* 226–244.

Bartz, J. A., & Lydon, J. E. (2006). Navigating the interdependence dilemma: Attachment goals and the use of communal norms with potential close others. *Journal of Personality and Social Psychology, 91,* 77–96.

Berscheid, E., & Regan, P. (2005). *The psychology of interpersonal relationships.* Upper Saddle River, NJ: Prentice Hall.

Birnbaum, G. E., Reis, H. T., Mikulincer, M., Gillath, O., & Orpaz, A. (2006). When sex is more than just sex: Attachment orientations, sexual experience, and relationship quality. *Journal of Personality and Social Psychology, 91,* 929–943.

Bookwala, J., & Zdaniuk, B. (1998). Adult attachment styles and aggressive behavior within dating relationships. *Journal of Social and Personal Relationships, 15*, 175–190.

Boon, S. D., & Griffin, D. W. (1996). The construction of risk in relationships: The role of framing in decisions about intimate relationships. *Personal Relationships, 3*, 293–306.

Bowlby, J. (1973). *Attachment and loss: Vol. 2. Separation: Anxiety and anger*. New York: Basic Books.

Bowlby, J. (1982). *Attachment and loss: Vol. 1. Attachment* (2nd ed.). New York: Basic Books. (Original manuscript published 1969.)

Bradford, S. A., Feeney, J. A., & Campbell, L. (2002). Links between attachment orientations and dispositional and diary-based measures of disclosure in dating couples: A study of actor and partner effects. *Personal Relationships, 9*, 491–506.

Brassard, A., Shaver, P. R., & Lussier, Y. (2007). Attachment, sexual experience, and sexual pressure in romantic relationships: A dyadic approach. *Personal Relationships, 14*, 475–494.

Brennan, K. A., Clark, C. L., & Shaver, P. R. (1998). Self-report measurement of adult attachment: An integrative overview. In J. A. Simpson & W. S. Rholes (Eds.), *Attachment theory and close relationships* (pp. 46–76). New York: Guilford Press.

Brennan, K. A., & Shaver, P. R. (1998). Attachment styles and personality disorders: Their connections to each other and to parental divorce, parental death, and perceptions of parental caregiving. *Journal of Personality, 66*, 835–878.

Cassidy, J., & Berlin, L. J. (1994). The insecure/ambivalent pattern of attachment: Theory and research. *Child Development, 65*, 971–991.

Cassidy, J., & Kobak, R. R. (1988). Avoidance and its relationship with other defensive processes. In J. Belsky & T. Nezworski (Eds.), *Clinical implications of attachment* (pp. 300–323). Hillsdale, NJ: Erlbaum.

Cassidy, J., & Shaver, P. R. (Eds.). (2008). *Handbook of attachment: Theory, research, and clinical applications* (2nd ed.). New York: Guilford Press.

Chen, E. C., & Mallinckrodt, B. (2002). Attachment, group attraction, and self-other agreement in interpersonal circumplex problems and perceptions of group members. *Group Dynamics, 6*, 311–324.

Christensen, A. (1988). Dysfunctional interaction patterns in couples. In P. Noller & M. A. Fitzpatrick (Eds.), *Perspectives on marital interaction* (pp. 31–52). Philadelphia: Multilingual Matters.

Collins, N. L., & Feeney, B. C. (2000). A safe haven: An attachment theory perspective on support seeking and caregiving in intimate relationships. *Journal of Personality and Social Psychology, 78*, 1053–1073.

Crawford, T. N., Livesley, W. J., Jang, K. L., Shaver, P. R., Cohen, P., & Ganiban, J. (2007). Insecure attachment and personality disorder: A twin study of adults. *European Journal of Personality, 21*, 191–208.

Crawford, T. N., Shaver, P. R., Cohen, P., Pilkonis, P. A., Gillath, O., & Kasen, S. (2006). Self-reported attachment, interpersonal aggression, and personality disorder in a prospective community sample of adolescents and adults. *Journal of Personality Disorders, 20*, 331–351.

Cyranowski, J. M., & Andersen, B. L. (1998). Schemas, sexuality, and romantic attachment. *Journal of Personality and Social Psychology, 74*, 1364–1379.

Davis, D., Shaver, P. R., & Vernon, M. L. (2003). Physical, emotional, and behavioral reactions to breaking up: The roles of gender, age, emotional involvement, and attachment style. *Personality and Social Psychology Bulletin, 29*, 871–884.

Davis, D., Shaver, P. R., & Vernon, M. L. (2004). Attachment style and subjective motivations for sex. *Personality and Social Psychology Bulletin, 30*, 1076–1090.

Donnellan, M. B., Burt, S. A., Levendosky, A. A., & Klump, K. L. (2008). Genes, personality, and attachment in adults: A multivariate behavioral genetic analysis. *Personality and Social Psychology Bulletin, 34*, 3–16.

Downey, G., & Feldman, S. I. (1996). Implications of rejection sensitivity for intimate relationships. *Journal of Personality and Social Psychology, 70*, 1327–1343.

Edelstein, R. S., & Shaver, P. R. (2004). Avoidant attachment: Exploration of an oxymoron. In D. Mashek & A. Aron (Eds.), *Handbook of closeness and intimacy* (pp. 397–412). Mahwah, NJ: Erlbaum.

Fraley, R. C., & Shaver, P. R. (1998). Airport separations: A naturalistic study of adult attachment dynamics in separating couples. *Journal of Personality and Social Psychology, 75*, 1198–1212.

Fraley, R. C., & Waller, N. G. (1998). Adult attachment patterns: A test of the typological model. In J. A. Simpson & W. S. Rholes (Eds.), *Attachment theory and close relationships* (pp. 77–114). New York: Guilford Press.

Gillath, O., Shaver, P. R., Baek, J., & Chun, D. S. (2008). Genetic correlates of adult attachment style. *Personality and Social Psychology Bulletin, 34*, 1396–1405.

Grabill, C. M., & Kerns, K. A. (2000). Attachment style and intimacy in friendship. *Personal Relationships, 7*, 363–378.

Hazan, C., & Shaver, P. R. (1987). Romantic love conceptualized as an attachment process. *Journal of Personality and Social Psychology, 52*, 511–524.

Hazan, C., & Zeifman, D. (1994). Sex and the psychological tether. In K. Bartholomew & D. Perlman (Eds.), *Advances in personal relationships: Attachment processes in adulthood* (Vol. 5, pp. 151–177). London: Jessica Kingsley.

Hazan, C., & Zeifman, D. (1999). Pair-bonds as attachments: Evaluating the evidence. In J. Cassidy & P. R. Shaver (Eds.), *Handbook of attachment: Theory, research, and clinical applications* (pp. 336–354). New York: Guilford Press.

Henderson, A. J. Z., Bartholomew, K., & Dutton, D. G. (1997). He loves me; he loves me not: Attachment and separation resolution of abused women. *Journal of Family Violence, 12*, 169–191.

Hesse, E. (2008). The Adult Attachment Interview: Protocol, method of analysis, and empirical studies. In J. Cassidy & P. R. Shaver (Eds.), *Handbook of attachment: Theory, research, and clinical applications* (2nd ed., pp. 552–598). New York: Guilford Press.

Horowitz, L. M., Rosenberg, S. E., Baer, B. A., Ureno, G., & Villasenor, V. (1988). Inventory of Interpersonal Problems: Psychometric properties and clinical applications. *Journal of Consulting and Clinical Psychology, 56*, 885–892.

Horppu, R., & Ikonen-Varila, M. (2001). Are attachment styles general interpersonal orientations? Applicants' perceptions and emotions in interaction with evaluators in a college entrance examination. *Journal of Social and Personal Relationships, 18*, 131–148.

Impett, E. A., & Peplau, L. A. (2002). Why some women consent to unwanted sex with a dating partner: Insights from attachment theory. *Psychology of Women Quarterly, 26*, 360–370.

Kirkpatrick, L. A., & Hazan, C. (1994). Attachment styles and close relationships: A four-year prospective study. *Personal Relationships, 1*, 123–142.

Kohut, H. (1977). *The restoration of the self.* New York: International Universities Press.

Lavy, S. (2006). *Expressions and consequences of intrusiveness in adult romantic relationships: An attachment theory perspective.* Unpublished doctoral dissertation, Bar-Ilan University, Ramat Gan, Israel.

Lavy, S., Mikulincer, M., & Shaver, P. R. (2010). Autonomy-proximity imbalance: An attachment theory perspective on intrusiveness in romantic relationships. *Personality and Individual Differences, 48*, 552–556.

Luborsky, L., & Crits-Christoph, P. (1998). *Understanding transference: The Core Conflictual Relationship Theme method.* Washington, DC: American Psychological Association.

Mikulincer, M. (1995). Attachment style and the mental representation of the self. *Journal of Personality and Social Psychology, 69*, 1203–1215.

Mikulincer, M., & Florian, V. (1998). The relationship between adult attachment styles and emotional and cognitive reactions to stressful events. In J. A. Simpson & W. S. Rholes (Eds.), *Attachment theory and close relationships* (pp. 143–165). New York: Guilford Press.

Mikulincer, M., Gillath, O., & Shaver, P. R. (2002). Activation of the attachment system in adulthood: Threat-related primes increase the accessibility of mental representations of attachment figures. *Journal of Personality and Social Psychology, 83*, 881–895.

Mikulincer, M., & Horesh, N. (1999). Adult attachment style and the perception of others: The role of projective mechanisms. *Journal of Personality and Social Psychology, 76*, 1022–1034.

Mikulincer, M., & Nachshon, O. (1991). Attachment styles and patterns of self-disclosure. *Journal of Personality and Social Psychology, 61*, 321–331.

Mikulincer, M., & Orbach, I. (1995). Attachment styles and repressive defensiveness: The accessibility and architecture of affective memories. *Journal of Personality and Social Psychology, 68*, 917–925.

Mikulincer, M., & Shaver, P. R. (2003). The attachment behavioral system in adulthood: Activation, psychodynamics, and interpersonal processes. In M. P. Zanna (Ed.), *Advances in experimental social psychology* (Vol. 35, pp. 53–152). New York: Academic Press.

Mikulincer, M., & Shaver, P. R. (2007a). *Attachment in adulthood: Structure, dynamics, and change.* New York: Guilford Press.

Mikulincer, M., & Shaver, P. R. (2007b). Boosting attachment security to promote mental health, prosocial values, and inter-group tolerance. *Psychological Inquiry, 18*, 139–156.

Mikulincer, M., Shaver, P. R., Bar-On, N., & Ein-Dor, T. (2010). The pushes and pulls of close relationships: Attachment insecurities and relational ambivalence. *Journal of Personality and Social Psychology, 98*, 450–468.

Mikulincer, M., & Sheffi, E. (2000). Adult attachment style and cognitive reactions to positive affect: A test of mental categorization and creative problem solving. *Motivation and Emotion, 24*, 149–174.

Morgan, H. J., & Shaver, P. R. (1999). Attachment processes and commitment to romantic relationships. In J. M. Adams & W. H. Jones (Eds.), *Handbook of interpersonal commitment and relationship stability* (pp. 109–124). New York: Plenum.

Ognibene, T. C., & Collins, N. L. (1998). Adult attachment styles, perceived social support, and coping strategies. *Journal of Social and Personal Relationships, 15*, 323–345.

Pierce, T., & Lydon, J. (1998). Priming relational schemas: Effects of contextually activated and chronically accessible interpersonal expectations on responses to a stressful event. *Journal of Personality and Social Psychology, 75*, 1441–1448.

Pistole, M. (1994). Adult attachment styles: Some thoughts on closeness-distance struggles. *Family Process, 33*, 147–159.

Raz, A. (2002). *Personality, core relationship themes, and interpersonal competence among young adults experiencing difficulties in establishing long-term relationships*. Unpublished doctoral dissertation, Haifa University, Haifa, Israel.

Reis, H. T., & Shaver, P. R. (1988). Intimacy as an interpersonal process. In S. Duck (Ed.), *Handbook of research in personal relationships* (pp. 367–389). London: Wiley.

Rholes, W. S., Simpson, J. A., Campbell, L., & Grich, J. (2001). Adult attachment and the transition to parenthood. *Journal of Personality and Social Psychology, 81*, 421–435.

Senchak, M., & Leonard, K. E. (1992). Attachment styles and marital adjustment among newlywed couples. *Journal of Social and Personal Relationships, 9*, 51–64.

Shaver, P. R., & Mikulincer, M. (2002). Attachment-related psychodynamics. *Attachment and Human Development, 4*, 133–161.

Shaver, P. R., Schachner, D. A., & Mikulincer, M. (2005). Attachment style, excessive reassurance seeking, relationship processes, and depression. *Personality and Social Psychology Bulletin, 31*, 343–359.

Simpson, J. A., Rholes, W. S., & Nelligan, J. S. (1992). Support seeking and support giving within couples in an anxiety-provoking situation: The role of attachment styles. *Journal of Personality and Social Psychology, 62*, 434–446.

Thompson, M. M., & Holmes, J. G. (1996). Ambivalence in close relationships: Conflicted cognitions as a catalyst for change. In R. M. Sorrentino & E. T. Higgins (Eds.), *Handbook of motivation and cognition*, Vol. 3: *The interpersonal context* (pp. 497–530). New York: Guilford Press.

Tracy, J. L., Shaver, P. R., Albino, A. W., & Cooper, M. L. (2003). Attachment styles and adolescent sexuality. In P. Florsheim (Ed.), *Adolescent romantic relations and sexual behavior: Theory, research, and practical implications* (pp. 137–159). Mahwah, NJ: Lawrence Erlbaum Associates.

Vogel, D. L., & Wei, M. (2005). Adult attachment and help-seeking intent: The mediating roles of psychological distress and perceived social support. *Journal of Counseling Psychology, 52*, 347–357.

CHAPTER 3

Outsourcing Effort to Close Others

GRÁINNE M. FITZSIMONS

University of Waterloo

ELI J. FINKEL

Northwestern University

Close relationship partners are highly interdependent. This interdependence is often obvious to others—most romantic couples' everyday lives are noticeably meshed or overlapped, as partners share everything from favorite television shows to social networks—but it can also exist in ways that are less tangible or visible to others. According to prominent models of interpersonal cognition (e.g., Aron, Aron, Tudor, & Nelson, 1991; Baldwin, 1992) and recent theorizing from social neuroscience (Coan, 2008), individuals within close relationships possess tightly associated and overlapping cognitive representations of self and other. Although most research on the effects of cognitive interdependence has emphasized its impact on the nature and quality of the relationships themselves, it is likely that these self–other links impact both individuals in important ways as well. Indeed, romantic partners' interdependence has been found to impact basic memory processes, such that partners come to rely on each other's memory for specific domains of knowledge and, as such, fail to retain information about those domains themselves (Wegner, Erber, & Raymond, 1991). According to this research, *transactive memory*, a system of shared encoding, storing, and retrieving information, allows romantic partners to specialize their own individual memories and together remember a greater amount than they would alone.

In the current chapter, we suggest that the same kinds of processes that encourage the development of transactive memory in close relationships may also extend to the domain of self-regulation (defined as the processes through which the self changes or alters itself to achieve goals; see Baumeister, Vohs, & Tice, 2007). In particular, we speculate that interdependence may lead to the development of *transactive self-regulation*, a system of shared or dyadic self-regulation that is stronger than the partners' individual systems of self-regulation. Just as partners may rely on each other's help to remember certain types of information, they may also rely on each other's help to make progress on certain goals. If partners rely on each other for self-regulatory help, their overall self-regulatory strength should be increased—by externalizing some of their efforts, they can save their own resources for other self-regulation challenges. However, relying on others can also leave individuals vulnerable to temporary self-regulatory failures, when circumstances promote reliance on the partner. It is this latter idea—that individuals outsource effort to helpful romantic partners—that we examine empirically in this chapter. Before we turn to that research, however, we begin by describing the conceptual background for our research. We first situate our findings within the broader literature on interpersonal effects on self-regulation.

Interpersonal Influences on Self-Regulation

In the field of social psychology, research has tended to describe self-regulation as an intrapersonal process, influenced by individual differences in self-regulatory styles and abilities and situational variables that promote the activation of different cognitive and motivational processes (for a review, see Baumeister, Schmeichel, & Vohs, 2007). For example, research has demonstrated that individuals are especially effective at self-regulation to the extent that they, for example, delay gratification (Mischel, 1974), use self-regulatory resources sparingly and strategically (Baumeister & Heatherton, 1996), believe that they have the requisite skill to achieve their goals (Bandura, 1986), and view setbacks as opportunities to learn (Dweck & Leggett, 1988).

In recent years, however, social psychological research has also documented diverse routes through which interpersonal processes and variables can impact self-regulation (for reviews, see Finkel & Fitzsimons, in press; Fitzsimons & Finkel, 2010). For example, research has demonstrated that individuals are especially effective at self-regulation to the extent that they think of a close relationship partner who values an active goal (Fitzsimons & Bargh, 2003; Shah, 2003). For example, after thinking about their mother, participants were more successful at an academic achievement task if they believed their mother valued their academic achievement

(Fitzsimons & Bargh). Individuals are also more effective at self-regulation to the extent that they have recently experienced efficient (i.e., smooth and easy) rather than inefficient (i.e., difficult, awkward, or challenging) social interactions (Finkelet al., 2006; Richeson & Trawalter, 2005; Vohs, Baumeister, & Ciarocco, 2005). Interpersonal processes have been shown to impact different types of self-regulatory processes, from the content of the goals pursued (Fitzsimons & Bargh) to the resources available to pursue those goals (Finkel et al.). The empirical work described in this chapter extends this burgeoning research on the effects of interpersonal processes on self-regulation by examining an interpersonal influence on the exertion of effort, another important component of self-regulation.

Because self-regulation is a limited resource (Baumeiste, Schmeichel, & Vohs, 2007), and because it is needed for the great majority of everyday goal pursuits, individuals will frequently tax and deplete that resource and will sometimes feel that they don't have sufficient self-regulatory resources to expend effort toward all of their important goals. For example, after a long and stressful day at the office, individuals can then be faced with a whole other set of challenges that require effort, including dealing with family obligations, household chores, and health and fitness goals. In part because of the limited nature of self-regulatory resources, many individuals will find it tough to persist successfully on all their goal pursuits. On an average night, average individuals will abandon healthy dinner plans, consume second helpings of dessert, ignore unpaid bills, and allow their children to watch extra television—not because these individuals don't have goals to eat healthy, do household chores, and spend quality time with their children but simply because they struggle with self-regulation. Indeed, research has consistently shown that all conscious acts of self-regulation draw upon the same limited resource and leave less of the resource available for subsequent acts of self-regulation (see Baumeister et al. for review). For example, research participants who have to resist eating delicious treats subsequently persisted for less time on unsolvable puzzles (Baumeister, Bratslavsky, Muraven, & Tice, 1998).

Given the limitations of intrapersonal resources of self-regulation, it seems likely that individuals may turn to their interpersonal environments for assistance with their goal pursuits. Indeed, research in both the close relationships and health subfields of psychology has found strong support for the important role that relationship partners can play in shaping self-regulation (Finkel & Fitzsimons, in press). For example, individuals who have romantic partners who are strongly supportive of their individual goal pursuits (e.g., in academics and fitness) feel more confident about their ability to achieve those goals and are ultimately more likely to achieve them than do individuals who have romantic partners who are less supportive (Brunstein, Dangelmayer, & Schultheiss, 1996; Feeney, 2004).

Although the mechanisms driving these effects of social support on successful self-regulation are not yet agreed upon, researchers have compellingly demonstrated that such effects are robust, at least when it comes to health goals (DiMatteo, 2004; Reblin & Uchino, 2008).

In a program of research on the *Michelangelo phenomenon*, Rusbult and colleagues have demonstrated one route through which close relationship partners can promote self-regulatory success. According to this research, relationship partners help to bring about each other's ideal self by acting as each other's "sculptors" (see Rusbult, Finkel, & Kumashiro, 2009, for review). Partners who see individuals as already possessing their ideal characteristics and who behave in ways that affirm those characteristics tend to promote or facilitate individuals' growth toward those ideal self-goals (Drigotas, Rusbult, Wieselquist, & Whitton, 1999).

Other research has demonstrated that, when a given goal is active, individuals tend to feel closer to and more readily approach relationship partners who are instrumental or helpful toward that goal's progress (Fitzsimons & Shah, 2008). These findings indirectly suggest that individuals may lean on their close relationship partners for help with ongoing goal pursuits, an interpretation further supported by a study demonstrating that individuals who draw closer to helpful others for active goals subsequently are more likely to succeed on those goals (Fitzsimons & Shah).

Thus, in addition to making individuals feel more positively about their relationships and more valuable and loved by their partners, supportive partners also help individuals achieve their goals (Brunstein et al., 1996). In the current work, we seek to extend these findings by investigating whether individuals come to rely on supportive partners for help with their goals. In other words, do individuals take advantage of having helpful partners? Do they conserve their own limited self-regulation resources and instead rely on their partners to help them with their ongoing goal pursuits? This work is a first step toward examining the possibility that, within close relationships, the self-regulation system is shared, just like the memory system has been shown to be shared (Wegner et al., 1991). If our research supports the idea that individuals outsource effort to close others, relying on those others to act as external agents of self-regulation, we will have some initial evidence for the existence of a transactive system of self-regulation within close relationships. Just as individuals fail to remember information that they typically rely on their partners to remember (Wegner et al., 1991), we hypothesize that individuals will fail to expend effort on a goal—that is, will reduce engagement of their individual self-regulation system—if they think about how they rely on their partners for help with that goal.

Current Research: Outsourcing Effort to Helpful Others

This chapter describes recent research that examines a novel phenomenon that we have termed *self-regulatory outsourcing*. By outsourcing, we mean that individuals exert less effort to achieve one of their goals after thinking about how their significant other may be instrumental for helping them achieve that goal. We suggest that when individuals think about how a partner can help with an ongoing goal, they unconsciously "outsource" the self-regulatory effort to their partner. By thinking about how helpful the partner can be, they then feel a greater sense of reliance on the partner for goal progress, and they thus feel less need to use their limited self-regulatory resources to pursue the goal themselves. Thus, they subsequently reduce their own effort. For example, imagine that Benjamin wants to eat less junk food. We suggest that if he thinks about how his wife, Carla, is ruthless about throwing out any candy or snacks around the house, he will feel less motivated to expend his own effort to avoid eating junk food because he will temporarily feel like he can rely on Carla. As another example, imagine that Carla wants to better control her stress. We suggest that if she thinks about how Benjamin calms her down, she will be less motivated to avoid potential triggers of stress because she will feel like she can rely on Benjamin.

The outsourcing hypothesis may seem to contradict past research on the benefits of having social support (e.g., Brunstein et al., 1996), which suggests that having supportive partners can increase self-efficacy and motivation. However, this hypothesis fits well with a number of well-established theories of effort exertion from the self-regulation field. For example, individuals have been shown to exert less effort when they feel that they are making satisfactory progress toward their goal (Carver & Scheier, 1990; Fishbach & Dhar, 2005). Perceiving that individuals have made progress is thought to generate positive affect and to reduce their motivation toward further expending effort. Individuals have also been shown to exert less effort when they are aware that others are also expending effort toward a common goal (Latané, Williams, & Harkins, 1979). That is, individuals are likelier to engage in reduced effort when *social loafing* is a viable route to goal achievement. Finally, individuals have been shown to exert less effort using any specific means to goal progress when they are aware of multiple means or routes to goal achievement (Kruglanski, Shah, Fishbach, Friedman, Chun, & Sleeth-Keppler, 2002). According to research on multiple means to goal attainment, the cognitive link between an overarching goal (e.g., weight loss) and any given means or subgoal (e.g., dieting) is weakened by the existence of links between the goal (weight loss) and other means (e.g., exercise). Potentially, then, thinking about the partner's effort (one means to goal attainment) can reduce people's strength of engagement with their original means (expending one's own effort). These well-established findings about effort and self-regulation all contribute possible mechanisms for why thinking about a helpful other might have a negative effect on effort expenditure.

Although at first glance this phenomenon may appear self-defeating, our assumption is that, in ongoing relationships, relying on the partner is a reasonable strategy that will often lead to greater overall self-regulatory success. It makes sense to reduce one's own effort expenditure if one's partner can achieve self-regulatory success on one's behalf, because then one can use those conserved resources toward another goal. Whether outsourcing in ongoing relationships is ultimately productive, of course, depends on a number of factors, most obviously the partner's reliability or responsiveness—will help be there in the future? In future research, we hope to explore these variables. In the context of our studies, however, we describe the effects of outsourcing as producing a negative effect, in that it undermines motivation to expend effort toward goal pursuit.

Potential Moderators of Self-Regulatory Outsourcing

In addition to seeking evidence for the basic outsourcing phenomenon, we also examine two potential moderators of this effect. First, we predict that individuals should be most likely to outsource effort when their own self-regulatory resources are low. As we have described, research on *depletion theory* has argued that all acts of self-regulation draw on one limited resource; tapping that resource by engaging in self-regulation depletes the resource, leaving less available energy for subsequent activities. Depletion is typically demonstrated by individuals' preference for easier tasks and reduced effort on any given task (for reviews, see Baumeister, Schmeichel, & Vohs, 2007; Muraven & Baumeister, 2000). We suggest that individuals who are depleted should be especially likely to show the outsourcing effect. They should be even more likely than their nondepleted counterparts to exhibit less effort after thinking about a helpful partner.

What about individuals who are not depleted but who are concerned that performing a current task will cause them to become depleted—will they reduce their resources for a subsequent task requiring self-regulatory resources? According to *resource conservation theory*, a recent offshoot of depletion theory, individuals are frequently motivated to conserve their limited resources for a subsequent self-regulatory task, and this motivation causes them to strategically limit resource expenditures on a current task (Muraven, Shmueli, & Burkley, 2006). We suggest that the outsourcing dynamics that undermine self-regulatory exertion in general are particularly powerful when conservation concerns are salient. Relative to individuals who believe that performing a current task will have no impact on their resources for the subsequent goal-relevant task, individuals who believe that performing a current task will deplete their resources for the subsequent task will conserve more resources for the subsequent task—unless they have outsourced their self-regulatory efforts to their romantic partner. Individuals who have done so should conserve relatively few self-regulatory resources for the subsequent task regardless of whether they perceive that performing the current task will diminish their available resources for the subsequent task.

The Consequences of Self-Regulatory Outsourcing for Relationship Commitment

In addition to seeking evidence for the basic outsourcing effect and to identifying potential moderators of theoretical relevance, we also seek to understand the potential relationship consequences of outsourcing to relationship partners. In particular, we examine the effects of outsourcing on commitment to the helpful partner. Although there is reason to believe that individuals will experience diminished commitment to a relationship partner who is currently undermining their goal-pursuit motivation, we hypothesize that the outsourcing process will increase individuals' relationship commitment. Based on prior work suggesting that individuals feel more positive about others who are instrumental for active goals (Fitzsimons & Shah, 2008), we speculate that when individuals think about how a partner is helpful for the achievement of their goals they will feel more gratitude and appreciation for the partner. In addition, the carefree quality that accompanies demotivation, because it is relaxed and effortless, can be pleasurable; thus, the combination of focusing on how a partner helps individuals achieve their goals while simultaneously experiencing the relaxation associated with self-regulatory outsourcing is likely to be associated with experiencing a particularly strong commitment to their partner. Finally, to the extent that individuals have heightened awareness of how they depend on their partner, they may feel a greater sense of dependence on the partner and thus feel more committed to maintaining the relationship over time.

Overview of the Hypotheses and Studies

We describe three experimental studies that test four related hypotheses stemming from our basic interest in the idea that individuals can outsource effort to their close others. Our primary hypothesis, the *outsourcing hypothesis*, states that individuals will exhibit diminished motivation toward a given goal when they think about ways a romantic partner helps them achieve that goal. Two of our hypotheses investigate moderators of the outsourcing effect. The *depletion hypothesis* states that the outsourcing effect should be especially pronounced among depleted individuals, whereas the *conservation hypothesis* states that it should be especially pronounced among individuals who believe that exertion on a current task can undermine performance on an important task to be performed moments later. Finally, the *relationship commitment hypothesis* states that, for individuals who have thought about how a romantic partner helps them achieve a given goal, greater outsourcing (i.e., greater reduction in individuals' own effort expenditure on the goal) will predict greater reported commitment to the relationship with this romantic partner.

Experiments 1 and 3 examine these hypotheses in the context of health and fitness goals. There are several reasons we chose to study outsourcing in the context of these specific goals. First, changes to health and fitness behaviors (e.g., eating more fruits and vegetables, cutting down on fat intake, starting a new exercise regimen) are commonly reported personal goals in most adult Western samples. Thus, we can assume that for most of our participants (adult females residing in the United States) this is a currently active goal. Second, the success or failure of health and fitness goals has profound consequences for individuals, perhaps even more so than do most career and financial goals. Failures to eat healthy and to exercise regularly are known to be major causes of physical and mental health problems in addition to strong predictors of premature morbidity (Finkelstein, Ruhm, & Kosa, 2005; Kopelman, 2000). Finally, health and fitness goals are, for many individuals, goals that require a great deal of self-control and effort to be successful. For example, it is widely known that most diets fail (Mann, Tomiyama, Westling, Lew, Samuels, & Chatman, 2007). Although there are many reasons diets tend to fail, one of the primary factors appears to be self-regulatory failure (Stroebe, 2008) because it is challenging for individuals to overcome well-established habits, to resist temptation, and to persist at such changes over time (Rothman, Sheeran, & Wood, 2009). In sum, health and fitness goals are common, important, and challenging and, as such, provide the ideal context for studying interpersonal effects on goal pursuit. Experiment 2 examines these hypotheses within the context of an academic achievement goal to establish that the effect generalizes beyond the health and fitness domain. Our sample for Experiment 2 is an undergraduate sample; thus, academic achievement goals possess the desired characteristics of being common, important, and challenging.

Empirical Tests of the Outsourcing Phenomenon, Moderators, and Consequences

Testing the Outsourcing and Depletion Hypotheses

In Experiment 1, we aimed to provide a first test of the outsourcing hypothesis—that individuals will expend less effort (in this case, that individuals will plan to expend less effort) toward a goal when they think about how their partner helps them achieve that goal. In addition to establishing the basic outsourcing phenomenon, Experiment 1 also tested the depletion hypothesis—that the outsourcing effect will be stronger under conditions of self-regulatory depletion. Experiment 1 employed a 2 × 2 design, with partner instrumentality (for a health/fitness goal vs. for a career goal) and depletion (low vs. high) as between-subjects variables. As a dependent measure, we assessed participants' willingness to exert effort to achieve their health/fitness goal (the focal goal). In addition, we assessed participants' commitment to their health/fitness goal. We have assumed that outsourcing diminishes the motivation to expend effort to pursue the focal goal without influencing the goal's importance. That is, individuals who outsource effort to a partner should still report caring about health/fitness and hoping to succeed. Outsourcing should change reliance on the partner but should not decrease the importance of the health/fitness goal.

Adult women, recruited from an online data collection service, completed the study online. They first completed a depletion manipulation modified from Muraven, Gagné, and Rosman (2008), in which they retyped a paragraph that appeared onscreen while skipping all vowels (*low depletion* condition) or while skipping all vowels that appeared two letters after another vowel (*high depletion* condition). Next, they provided one example of how their partner made it easier for them to do better with their health and fitness goals (focal goal condition) or career goals (control goal condition). Participants then indicated how much time and effort they planned to spend on their health and fitness goals in the upcoming week, using simple one-item Likert measures (1= much less time (or effort) than usual; 5 = much more time (or effort) than usual), and reported their commitment to health and fitness using a two-item scale that asked participants to rate their agreement (1 = I completely disagree; 7 = I completely agree) with the statements, "My health and fitness goals are important to me," and "I care about my progress on my health and fitness goals."

Results supported our hypotheses. As predicted, participants who thought about their romantic partner's instrumentality for their health and fitness goals planned to spend less time and effort pursuing health and fitness than did participants who thought about their partner's instrumentality for their career goals. Importantly, individuals in both conditions reported approximately equal levels of commitment to their health and fitness goals, suggesting that the outsourcing effect on effort expenditure is not driven by a reduction in the importance of the goal. In addition, as predicted, the effect was significantly stronger in the high depletion condition, suggesting that outsourcing may be more likely to occur when individuals have fewer available self-regulatory resources (Muraven et al., 2006). Thus, this first study provided support for both the outsourcing and depletion hypotheses.

Testing the Outsourcing and Conservation Hypotheses

In Experiment 2, we aimed to provide an additional test of the outsourcing hypothesis with a different sample of participants (undergraduate students) and in a different goal domain (academic achievement). We also sought to provide an initial test of the conservation hypothesis—that is, that participants who think about how their partner helps them achieve a given goal will conserve fewer resources for an anticipated goal-relevant task (compared with participants who think about how their partner helps them achieve a different goal). Experiment 2 employed a behavioral measure of effort expenditure instead of the self-report measure of planned effort expenditure that we used in Experiment 1 and included an additional control condition in which participants thought about something they like about their partner. This condition, which should produce positive feelings toward the partner, does not mention helpfulness or goals, and it provided a comparison for the two goal conditions. Experiment 2 employed a 3 × 2 design, with instrumentality condition (the focal goal of academic achievement, the control goal of recreation, and the control nongoal) and task framing (nondepleting, depleting) as between-subjects factors and time spent on distracter task as the dependent measure.

To test the conservation hypothesis, we adapted a procedure from research on the conservation model of self-regulation (Muraven et al., 2006). We manipulated participants' perception of whether an appealing distracter task consumed resources needed for a subsequent task that was strongly relevant for participants' academic achievement goals. Based on research on the conservation model of self-control (Muraven et al.), we used time spent on the first, distracter task as a measure of how much participants were trying to conserve resources for the second, target task—with more time spent on the distracter task indicating diminished emphasis on resource conservation. We predicted that participants would conserve fewer resources for the subsequent, target task (i.e., they would spend more time on the initial, distracter task) when they had thought about how their partner helps them with their academic achievement goal. We also predicted that this pattern would be strongest when participants believed that the two tasks competed for resources (i.e., that the first task would drain resources needed to complete the second target task).

Undergraduate men and women completed the study online. Participants first completed the instrumentality manipulation, a slightly modified version of Experiment 1's manipulation in which participants provided an example of how their romantic partner helped with their ongoing academic achievement goals (the focal goal condition) or an ongoing recreational goal (the control goal condition) or simply reported something that they liked about their partner (the control nongoal condition). Next, participants learned that that they would first complete an entertaining puzzle task and then a difficult academic task that would teach them skills to improve their multiple-choice test-taking performance. Participants learned that they could divide the remaining study time on the two tasks in whatever fashion they wished. Participants read instructions indicating that spending time on the first task would either drain their resources and make the second task harder and less useful (depleting frame condition) or that it would not drain their resources for the second task (non-depleting frame condition). Participants then spent as much time as they wished on the first task, which consisted of a number of easy word puzzles typically rated as enjoyable in prior research with a similar sample. There was no second task; whenever participants indicated they wished to continue, they received debriefing feedback. If participants reached 7 minutes, the program automatically ended and brought up the debriefing feedback.

Results supported our hypotheses. As predicted, participants spent significantly more time on the distracter task in the focal goal condition than in either of the other two conditions, which did not differ from each other. That is, thinking about how their romantic partner helped with an academic achievement goal led participants to conserve less effort for the academic task (i.e., to expend more energy on the first task), compared with thinking about how their romantic partner helped with another goal or with thinking about something nice about their romantic partner. Importantly, and as predicted, this pattern was significant only when participants thought the first task interfered with the second task. When the two tasks were not described as competing for resources, participants were unaffected by thinking about an instrumental romantic partner.

Testing the Outsourcing and Relationship Commitment Hypotheses

In Experiment 3, we attempted to replicate the basic outsourcing phenomenon and to test our relationship commitment hypothesis. This study followed the procedures of Experiment 1 but did not manipulate depletion condition, added a partner positivity control condition like that in Experiment 2, and, most importantly, assessed relationship commitment at the end of the session. As explained earlier in the chapter, we expected that outsourcing effort to a romantic partner would promote self-reported commitment to that partner.

As in Experiment 1, adult women recruited from an online data collection service completed the study online. They first typed in one example of how their partner helped with their current health and fitness goals (focal goal condition), one example of how their partner helped with their current career goals (control goal condition), or one thing they liked about their romantic partner (control nongoal). Next, they reported how much time they planned to spend pursuing their health and fitness goals in the upcoming week and completed a two-item relationship commitment measure, which asked them to indicate their agreement with two statements ("I am highly committed to my current partner," and "I believe I will stay with this partner for the rest of my life") on seven-point Likert scales.

Results supported both the outsourcing and relationship commitment hypotheses. Participants reported planning to spend significantly less time pursuing their health and fitness goals in the focal goal condition (i.e., after thinking about how their romantic partner helps with their health and fitness goals) than in either of the other two conditions (i.e., after thinking about how their partner helps with their career goal or about something they like about their partner). In addition, as predicted, participants' planned goal pursuit (i.e., how much time they reported they would spend on health and fitness in the upcoming week) was significantly negatively related to relationship commitment in the focal goal condition, but not in the two control conditions. We predicted, and found, a different relationship between planned goal pursuit and relationship commitment; we did not predict (nor find) a main effect of condition on the commitment measure. We assume that the positive effects of outsourcing processes did not bolster commitment above the level of the other two conditions, because there were other unique positive processes at work in those conditions as well. For example, the help partners gave for career goals was generally bigger in scale and highly positive—participants indicated examples like, "He supports who I am in life," and "He stands by me when I need emotional help," as it was in the positive control condition—participants wrote examples like, "He is the love of my life," and "He makes me a better woman." In contrast, the help partners gave for fitness goals was smaller and sometimes not entirely positive—participants gave examples like, "He watches the baby so I can get to the gym," and "He gives me a bad look when I eat too much." Thus, any main effect of outsourcing processes was likely overwhelmed by these between-goal differences.

Importantly, though, we found the predicted link between outsourcing and commitment in the focal goal condition. In other words, among women who thought about how their partner helps them achieve their health and fitness goals, greater outsourcing (i.e., greater reduction in the women's personal motivation to work hard to achieve the goal) predicted greater relationship commitment.

This latter finding provides some initial support for the idea that outsourcing effort to a romantic partner may generate positive outcomes for the relationship. Although outsourcing was not experimentally manipulated—that is, we did not induce some women to outsource more or less than others—and thus, we cannot say with certainty that outsourcing caused the increase in commitment, the data suggest that benefits may exist for the relationship. We speculate that outsourcing reflects positive qualities of the relationship, as it likely indicates greater reliance and trust in the partner's ability or willingness to help with personal goals. We further speculate that outsourcing may promote positive responses to partners, as it may remind individuals of their dependence on their partner. If the relationship were to end, the individuals would lose this external source of help with goal pursuit and thus may be more motivated to maintain the relationship over time.

General Discussion

In this chapter, we discussed several recent studies that examine a novel phenomenon that we have termed *self-regulatory outsourcing,* in which individuals exert less effort to achieve a goal after thinking about how a close other helps them to make progress on that goal. We speculate that one reason that the outsourcing effect occurs is that individuals tend to seek to conserve their limited self-regulatory resources, and, thus, when they think about how their partner helps with a goal they relax their own efforts, as a way of conserving resources. This idea is supported by our findings that the outsourcing effect is especially strong when individuals' resources have recently been depleted or when the need to conserve resources is made salient. In addition to seeking evidence for the basic effect and testing these moderators, we also examined the consequences of outsourcing for relationships. Although thinking of a helpful other appears to temporarily undermine individuals' motivation, and thus could potentially cause relationship strife, we hypothesized and found that outsourcing seems to predict positive relationship outcomes, at least in terms of commitment.

The most pressing direction for future research is to examine the possibility that outsourcing serves to promote engagement in other goals. If outsourcing is motivated by a desire to conserve resources, it may allow individuals to invest more resources in other goal pursuits. For example, outsourcing effort to a partner for progress on a health goal may allow an individual to use the conserved resources to pursue an ongoing career goal. This pattern of results would suggest that outsourcing does not have a negative impact on overall self-regulatory success. Indeed, if outsourcing reflects the operation of a broader shared system of self-regulation, then it may well maximize overall success for partners. To test for effects on other goals, future research could add a second task, one relevant to another important goal, to the paradigm used in the current studies. For example, after thinking about how a partner helps with health and fitness (vs. control) and having an opportunity to expend effort on a health and fitness–relevant task, participants would complete a task relevant to academic achievement. We would examine persistence and performance on the academic achievement task as a measure of motivation to expend effort on that goal. In addition to experimental data, it may also be useful to examine how naturally occurring outsourcing within a relationship affects progress on goals and perceived self-regulatory success over time.

We have discussed the possibility that outsourcing effort to a romantic partner may have positive consequences for relationships because it encourages reliance on the partner and may lead to greater feelings of appreciation of and gratitude to the partner. However, outsourcing likely varies in both occurrence and consequence—that is, it may be likelier to occur within some relationships and for some individuals, and it may have positive consequences for some relationships and negative consequences for others. For example, individuals who are less comfortable with dependence on a partner, such as those who are low in self-esteem, may be less likely to rely on a romantic partner for help with important goals. Individuals who are high in attachment avoidance may similarly be less likely to rely on a partner and may also experience discomfort and seek interpersonal distance when forced to think about reliance on a partner. Qualities of the partner will also likely play a moderating role on the occurrence and relationship consequences of outsourcing. For example, partners who encourage dependence (Feeney, 2004) and partners who are highly committed may both be likely to promote increased outsourcing.

Finally, even if outsourcing has benefits for the goal pursuer, there are aspects of the relationship or situation that might have costs for the partner who provides help. For example, imbalances in support for each other's goals could create burdens on the undersupported partner. If one partner consistently relies on the other for help and support, it may reduce the helpful partner's time and motivation to pursue his or her own goals, and the imbalance may also build resentment in some partners. In a committed long-term relationship, these imbalances may ebb and flow, such that one partner is alternately the primary "outsourcer" and the primary "outsourcee" across time, or they may become quite stable, with one partner being increasingly dependent on the other for help over time. In general, future research should complement the current focus on the recipient of outsourcing help with a focus on the provider of help and the dyadic effects of these kinds of self-regulatory processes. Future research should also include additional measures of relationship quality and functioning as well as measures of relevant individual differences to determine if they affect outsourcing dynamics as studied experimentally and should also investigate how outsourcing impacts relationship quality over time.

As we discussed in the introduction, research has suggested that romantic couples rely on each other's memories and eventually develop a shared memory structure (Wegner et al., 1991). In future research, we hope to extend the current findings about outsourcing effort to romantic partners to examine the broader idea that couples may develop shared self-regulatory systems, or *transactive self-regulation*, in which each partner relies on the other for help with their self-regulation tasks. Individuals who rely on their partner for help with self-regulation in one area or with some specific types of self-regulation problems may conserve valuable resources for other self-regulation efforts. If so, despite producing ironically negative short-term effects like the outsourcing effects shown here, such a shared self-regulatory system may ultimately serve to benefit partners if it allows them to best make use of their limited self-regulatory resources over time.

References

Aron, A., Aron, E. N., Tudor, M., & Nelson, G. (1991). Close relationships as including other in the self. *Journal of Personality and Social Psychology, 60*, 241–253.

Baldwin, M. W. (1992). Relational schemas and the processing of social information. *Psychological Bulletin, 112*, 461–484.

Bandura, A. (1986). *Social foundations of thought and action: A social cognitive theory*. Englewood Cliffs, NJ: Prentice-Hall.

Baumeister, R. F., Bratslavsky, E., Muraven, M., & Tice, D. M. (1998). Ego-depletion: Is the active self a limited resource? *Journal of Personality and Social Psychology, 74*, 1252–1265.

Baumeister, R. F., & Heatherton, T. F. (1996). Self-regulation failure: An overview. *Psychological Inquiry, 7*, 1–15.

Baumeister, R. F., Schmeichel, B. J., & Vohs, K. D. (2007). Self-regulation and the executive function: The self as controlling agent. In A. W. Kruglanski & E. T. Higgins (Eds.), *Social psychology: Handbook of basic principles* (2nd ed., pp. 516–539). New York: Guilford.

Baumeister, R. F., Vohs, K. D., & Tice, D. M. (2007). The strength model of self-control. *Current Directions in Psychological Science, 16*, 351–355.

Brunstein, J. C., Dangelmayer, G., & Schultheiss, O. C. (1996). Personal goals and social support in close relationships: Effects on relationship mood and marital satisfaction. *Journal of Personality and Social Psychology, 71*, 1006–1019.

Carver, C. S., & Scheier, M. F. (1990). Origins and functions of positive and negative affect: A control-process view. *Psychological Review, 97*, 19–35.

Coan, J. A. (2008). Toward a neuroscience of attachment. In J. Cassidy and P. R. Shaver (Eds.), *Handbook of attachment: Theory, research, and clinical applications* (2nd ed., pp. 241–265). New York: Guilford Press.

DiMatteo, M. R. (2004). Social support and patient adherence to medical treatment: A meta-analysis. *Health Psychology, 23*, 207–218.

Drigotas, S. M., Rusbult, C. E., Wieselquist, J., & Whitton, S. (1999). Close partner as sculptor of the ideal self: Behavioral affirmation and the Michelangelo phenomenon. *Journal of Personality and Social Psychology, 77*, 293–323.

Dweck, C. S., & Leggett, E. L. (1988). A social-cognitive approach to motivation and personality. *Psychological Review, 95*, 256–273.

Feeney, B. C. (2004). A secure base: Responsive support of goal strivings and exploration in adult intimate relationships. *Journal of Personality and Social Psychology, 87*, 631–648.

Finkel, E. J., Campbell, W. K., Brunell, A. B., Dalton, A. N., Chartrand, T. L., & Scarbeck, S. J. (2006). High-maintenance interaction: Inefficient social coordination impairs self-regulation. *Journal of Personality and Social Psychology, 91*, 456–475.

Finkel, E.J., & Fitzsimons, G. M. (in press). Effects of interpersonal relationships on self-regulation. In K. D. Vohs & R. F. Baumeister (Eds.), *Handbook of self-regulation*, 2nd ed.

Finkelstein, E. A., Ruhm, C. J., & Kosa, K. M. (2005). Economic causes and consequences of obesity. *Annual Review of Public Health, 26*, 239–257.

Fishbach, A., & Dhar, R. (2005). Goals as excuses or guides: The liberating effect of perceived goal progress on choice. *Journal of Consumer Research, 32*, 370–377.

Fishbach, A., Friedman, R. S., & Kruglanski, A. W. (2003). Leading us not unto temptation: Momentary allurements elicit overriding goal activation. *Journal of Personality and Social Psychology, 84*, 296–309.

Fitzsimons, G. M., & Bargh, J. A. (2003). Thinking of you: Nonconscious pursuit of interpersonal goals associated with relationship partners. *Journal of Personality and Social Psychology, 84*, 148–164.

Fitzsimons, G. M., & Finkel, E. J. (2010). Interpersonal influences on self-regulation. *Current Directions in Psychological Science, 19*, 101–105.

Fitzsimons, G. M., & Shah, J. Y. (2008). How goal instrumentality shapes relationship evaluations. *Journal of Personality & Social Psychology, 95*, 319–337.

Kopelman, P. G. (2000). Obesity as a medical problem. *Nature, 404*, 635–643.

Kruglanski, A. W., Shah, J. Y., Fishbach, A., Friedman, R., Chun, W. Y., & Sleeth-Keppler, D. (2002). A theory of goal systems. In M. P. Zanna (Ed.), *Advances in experimental social psychology* (Vol. 34, pp. 331–378). San Diego, CA: Academic Press.

Latané, B., Williams, K., & Harkins, S. (1979). Many hands make light the work: The causes and consequences of social loafing. *Journal of Personality & Social Psychology, 37*, 822–832.

Mann, T., Tomiyama, A. J., Westling, E., Lew, A., Samuels, B., & Chatman, J. (2007). Medicare's search for effective obesity treatments: Diets are not the answer. *American Psychologist, 62*, 220–233.

Mischel, W. (1974). Processes in delay of gratification. In L. Berkowitz (Ed.), *Advances in experimental social psychology* (Vol. 7, pp. 249–292). New York: Academic.

Muraven, M., & Baumeister, R. F. (2000). Self-regulation and depletion of limited resources: Does self-control resemble a muscle? *Psychological Bulletin, 126*, 247–259.

Muraven, M., Gagné, M., & Rosman, H. (2008). Helpful self-control: Autonomy support, vitality, and depletion. *Journal of Experimental Social Psychology, 44*, 573–585.

Muraven, M., Shmueli, D., & Burkley, E. (2006). Conserving self-control strength. *Journal of Personality and Social Psychology, 91*, 524–537.

Murray, S. L., & Holmes, J. G. (2008). The commitment insurance system: Self-esteem and the regulation of connection in close relationships. In M. P. Zanna (Ed.), *Advances in experimental social psychology* (Vol. 40, pp. 1–60). San Diego, CA: Academic Press.

Reblin, M., & Uchino, B. N. (2008). Social and emotional support and its implication for health. *Current Opinion in Psychiatry, 21*, 201–205.

Richeson, J. A., & Trawalter, S. (2005). Why do interracial interactions impair executive function? A resource depletion account. *Journal of Personality and Social Psychology, 88*, 934–947.

Rothman, A., Sheeran, P., & Wood, W. (2009). Reflective and automatic processes in the initiation and maintenance of dietary change. *Annals of Behavioural Medicine, 38*, S4–17.

Rusbult, C. E., Finkel, E. J., & Kumashiro, M. (2009). The Michelangelo phenomenon. *Current Directions in Psychological Science, 18*, 305–309.

Stroebe, W. (2008). *Dieting, overweight, and obesity: Self-regulation in a food-rich environment*. Washington, DC: American Psychological Association.

Vohs, K. D., Baumeister, R. F., & Ciarocco, N. (2005). Self-regulation and self-presentation: Regulatory resource depletion impairs impression management and effortful self-presentation depletes regulatory resources. *Journal of Personality and Social Psychology, 88*, 632–657.

Wegner, D. M., Erber, R., & Raymond, P. (1991). Transactive memory in close relationships. *Journal of Personality and Social Psychology, 61*, 923–929.

CHAPTER 4

Approaching Rewards and Avoiding Threats in Close Relationships

SHELLY L. GABLE

University of California–Santa Barbara

Close relationships can be the source of joy as well as the source of pain. The potential rewards of close relationships, such as companionship, passion, and intimacy, are valuable social incentives. The potential pitfalls, such as betrayal, jealousy, and criticism, are important social threats. Reflecting this reality, people are motivated to both obtain the incentives in relationships as well as avoid the threats. Some people seem to be successful in avoiding the threats in relationships, but they do not seem to be adept at obtaining the incentives. On the other hand, some people seem to be doing just fine in the incentive category but are have difficulty avoiding the threats. In the following sections, evidence is presented for the fundamental assumption that close relationships expose people to incentives as well as threats, and regulating both is crucial. In subsequent sections a model of incentive- and threat-based relationship motivation and behavior regulation is described and evidence in support of the model is presented.

Threats in Close Relationships

Many theorists and researchers have focused on understanding the impact of threats in close relationship. The weight of the evidence is compelling that aversive experiences in relationships have large physical and psychological consequences. For example, troubled relationships are the reason people most often give when seeking psychotherapy (e.g., Pinsker, Nepps, Redfield, & Winston, 1985), and relationship conflict and hostility are associated with physical reactions that are associated with increased risk for disease (e.g., Kiecolt-Glaser, 1999; Uchino, Holt-Lunstad, Uno, & Flinders, 2001). Researchers have explored how these important potential threats in close relationships regulate behavior.

Growing out of Bowlby's (1969) theory of infant attachment processes, many adult attachment researchers have focused on understanding how variations in the sensitivity to the threat of abandonment regulates behavior in close relationships (e.g., Collins & Feeney, 2000; Davis, Shaver, & Vernon, 2004). Individual differences in this attachment dimension referred to as abandonment anxiety have been linked to several important relationship processes, including cognitive processing (Mikulincer, 1998), relationship establishment (Feeney, Cassidy, & Ramos-Marcuse, 2008), relationship maintenance behaviors (Tran & Simpson, 2009), and caregiving (Collins & Feeney, 2000). Another approach to understanding differences in threat regulation has been work by Downey and colleagues on rejection sensitivity (Downey, Freitas, Michaelis, & Khouri, 1998).

Rejection sensitivity is the tendency to anxiously expect and strongly react to rejection or perceived rejection by close others. Similar to attachment style findings, variations in people's rejection sensitivity have been associated with close relationship processes, including behavior during problem-solving discussions, reported negative emotions after conflicts with a romantic partner, and even physical violence in relationships (Ayduk et al., 1999; Downey, Feldman, & Ayduk, 2000; Downey et al., 1998). Regulating hostility and negative emotion during interactions also has significant intrapersonal and interpersonal consequences (e.g., Gottman, 1994). For example, Baron, Smith, Butner, Nealy-Moore, Hawkins, and Uchino (2007) found that marital adjustment was negatively correlated with current anger and hostility ratings. Moreover, reports of hostility and anger predicted changes in marital satisfaction and ratings of conflict down the road.

In short, relationship threats such as abandonment, rejection, and conflict have important outcomes. There is also growing evidence that we are biologically prepared to regulate our behavior around such potential relationship threats. For example, using functional magnetic resonance imaging (fMRI) researchers have shown that rejection (Eisenberger, Lieberman, & Williams, 2003), social isolation (Eisenberger, Jarcho, Lieberman, & Naliboff, 2006), or even mentally picturing a conflict or breakup with a romantic partner (Gillath, Bunge, Shaver, Wendelken, & Mikulincer, 2005) have been associated with activation in the same areas of the brain that are active when people report experiencing physical pain (areas such as the dorsal anterior cingulate cortex). In addition to central nervous system involvement in response to relationship threats, other studies have shown that sympathetic activation and neuroendocrine reactivity occur during interpersonal conflict or threats of rejection (e.g., Kiecolt-Glaser, 1999). And work by Dickerson and colleagues on immune responses to social threats has shown that the threat of social devaluation and rejection can trigger a physiological stress response (Dickerson & Kemeny, 2004). This physiological stress response includes increased proinflammatory cytokine activity (Dickerson, Gable, Irwin, Aziz, & Kemeny, 2009) and a heightened cortisol stress response (Dickerson & Kemeny, 2004).

The mounting behavioral and biological evidence supports the idea that social threats have psychological and physical consequences and individuals regulate their behavior in an attempt to avoid these threats. It seems the most efficient way to avoid the threats inherent in the establishment and maintenance of close relationships would be to forego relationships. However, a large body of literature exists suggesting that social isolation is not a healthy option. For example, large-scale epidemiological work has shown that a lack of social connection is associated with increased risk for mortality (e.g., House, Landis, & Umberson, 1988). More recently, Cacioppo and colleagues showed that there are numerous psychological and physical risks associated with being chronically lonely (see Cacioppo & Patrick, 2008). Thus, relationships must provide vital benefits to individuals; this review now turns to the possible incentives or rewards that close relationships provide.

Incentives in Relationships

Although close relationships research has largely been considered with potential threats, there has been important work on the potential incentives of close relationships. One potential incentive of close relationships is positive emotion. Recent research on positive emotions has, contrary to previous assumptions, clearly shown that positive emotions are themselves important to health and well-being (for review see Lyubomirsky, King, & Diener, 2005). For example, displays of positive emotions by individuals at a young age predict mortality and well-being in later life (Danner, Snowdon, & Friesen, 2001; Harker & Keltner, 2001). A possible incentive provided by close relationships is the experience of positive emotions such as fun and amusement. Consistent with this idea, people routinely report that close relationships are sources of the most meaningful and joyful aspects of their lives (Klinger, 1977). Also consistent with this idea is that positive emotions and experiences seem to be integral to close relationship functioning. Art Aron and colleagues' research has long highlighted the importance of shared fun and recreation in close relationships. For example, Aron, Norman, Aron, McKenna, and Heyman (2000) found that participation in joint novel, arousing activities led to increased relationship satisfaction. Specifically, in a series of experiments, spouses completed novel tasks together (e.g., an obstacle course) and reported greater relationship satisfaction than participants who did these activities separately.

Recently, work by Algoe and colleagues has focused on the positive emotion of gratitude in close relationships (e.g., Algoe, Gable, & Maisel, 2010). Specifically, this work has shown that the emotion of gratitude plays an important role in relationship formation and maintenance and has offered evidence for this in several studies employing different methods and involving diverse samples (e.g., Algoe & Haidt, 2009; Algoe, Haidt, & Gable, 2008). Most relevant to the current discussion is a recent study of cohabiting couples in which both members of the dyad reported on their emotions and behaviors in a daily experience study (Algoe et al., 2010). Participants' emotions of gratitude predicted increases in relationship connection and satisfaction the following day, for both them and their partners (i.e., both the recipient and benefactor reported greater relationship quality). Evidence that positive emotions are important incentives in close relationships comes from the other side of coin as well. For example, the absence of positive emotions and of jointly experienced positive events (boredom) predicts marital distress down the road (Tsapelas, Aron, & Orbuch, 2009).

Some have suggested that the role positive emotions and other incentives play in ongoing relationships is limited and that managing potential threats such as conflict and the regulation of negative emotions is at the crux of close relationships (e.g., Notarius & Markman, 1993). However, recent longitudinal work has shown that incentives-based relationship behavior and processes play an important role in the maintenance of close bonds. For example, expressions of affection and love in newlyweds were negatively correlated with divorce 13 years later (Huston, Caughlin, Houts, Smith, & George, 2001). Thus, it seems that positive emotions and other incentives such as companionship, affection, and intimacy are important potential rewards of close relationships. However, is there any evidence that people are biologically predisposed to seek and react to social incentives?

Much of the research into the biology of social processes has focused on social threats; however, several prominent models outline the biological routes of social behavior include the existence of social or attachment system that regulate social rewards (e.g., Insel, 2000; Panksepp, 1998). For example, recent work on the role of oxytocin in both animals and humans suggests that it is released in response to social incentives (Carter, 1998; Taylor, Klein, Lewis, Gruenewald, Gurung, & Updegraff, 2000). For example, several researchers have found that positive social interactions, especially those entailing affectionate contact, are associated with increases in oxytocin level (Carter, 1998). In turn, oxytocin may also stimulate affectionate contact, affiliation (Insel, 2000), nurturance (Taylor et al., 2000), and trust (Zak, Kurzban, & Matzner, 2005). And, parallel to the work on social pain and physical pain previously reviewed, investigations using functional neuroimaging suggest that socially rewarding experiences activate brain regions associated with pleasure (e.g., Bartels & Zeki, 2000, 2004).

In summary, there is growing evidence suggesting that humans are biologically wired to seek social incentives, and close relationships provide the main pathway to those incentives. The literature also suggests that social incentives are integral to close relationship functioning. As reviewed in the previous section, there is good evidence that people are built to be sensitive to and react to social threats. Close relationships are the main sources of these threats, which in turn are detrimental to close relationship functioning. Thus, it seems apparent that people are both motivated to approach incentives and avoid threat in close relationships and that these motivational systems are likely independent of one another. In the remainder of this chapter, a model of approach and avoidance motives and goals in close relationships is presented, and the evidence supporting this model to date is reviewed.

Approach and Avoidance Motivation in Close Relationships

Although a large body of evidence supports the idea that across the life span people are strongly motivated to form and maintain close relationships (e.g., Baumeister & Leary, 1995; Reis, Collins, & Berscheid, 2000), close relationship processes have rarely been examined from a motivational or goal theory perspective. My colleagues and I have conducted empirical research and proposed a model outlining processes associated with the establishment, maintenance, and dissolution of close relationships that is based on motivation and goal theory (Gable, 2006; Gable & Berkman, 2008; Gable & Poore, 2008; Impett, Gordon, Kogan, Oveis, Gable, & Keltner, 2010). Our model and corresponding research highlights a critical dimension of motivation and behavior regulation—the focus of goals. Relationship goals can be focused on a rewarding or desired end-state (i.e., approach), or relationship goals can be focused on a punishing, undesired end-state (i.e., avoidance; e.g., Gable, 2006). Previous work on close relationships has been concerned with either the rewarding aspects (e.g., attraction, intimacy) or the punishing aspects (e.g., insecurity, conflict) but rarely has examined them in tandem. However, as reviewed in the previous sections, because close relationships offer both threats and incentives, it has been our contention that models of close relationship functioning are incomplete unless both approach and avoidance dimensions are considered. More importantly, our work has shown that the approach–avoidance distinction has important implications for attention, cognition, emotions, and behavior in the relationship context.

Motivation scholars have long recognized the importance of the approach–avoidance distinction in work on achievement, power, and affiliation (e.g., Atkinson, 1958; Miller, 1959; Schneirla, 1959). However, less attention has been given to the approach–avoidance motivational distinction in close relationships, despite the fact that scholars have long recognized both the costs and rewards of close relationships (for reviews see Gable & Reis, 2001; Reis & Gable, 2003). Although not directly focused on close relationships, early work by Atkinson, Heyns, and Veroff (1954), Boyatzis (1973), and Mehrabian (1976) examined both incentive-focused social motivation (e.g., need for affiliation) and threat-focused social motivation (e.g., fear of rejection). McAdams and colleagues conducted research on intimacy motivation, which they viewed as motivation more applicable to reciprocal and ongoing relationships (e.g., McAdams, Healy, & Krause, 1984), which is more aligned with our interest in close relationships; however, no distinction was made between approach or incentive-based intimacy motives and avoidance or threat-based intimacy motives in this research.

Model of Approach and Avoidance Goals in Close Relationships

Gable (2006) proposed an approach–avoidance model of social motivation to understand close relationship processes. This model grew out of other hierarchical models of approach–avoidance motivation (e.g., Elliot, 2006). As depicted in Figure 4.1, the model posits that individual differences in dispositional threat and incentive sensitivities (e.g., relatively stable traits) and the social environment (e.g., recent events in a particular close relationship) influence the goals individuals adopt approach and avoidance goals. Specifically, approach social motives (e.g., hope for affiliation) and the potential for incentives in the current social environment combine to predict whether people adopt short-term approach relationship goals, and avoidance social motives (e.g., fear of rejection) and the potential for

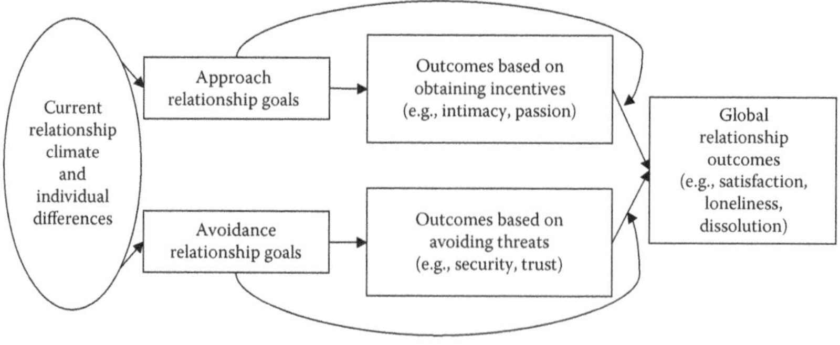

Figure 4.1 Approach–avoidance relationship goals and relationship outcomes.

current threats in the social environment combine to predict whether people will adopt avoidance relationship goals. Consistent with this idea, three studies by Gable (2006) found that individual differences in distal motives predicted more proximal goals. Specifically, individuals with strong approach motives were more likely to adopt short-term approach social goals (e.g., "Make new friends") and those with strong avoidance motives were more likely to adopt short-term avoidance social goals ("Not be lonely").

For example, in a newly established romantic relationship, Joe, a person who has strong approach social motives or who has had several exciting and fun dates with Chris, is likely to adopt approach goals regarding intimacy, such as, "I want to get closer and more intimate with Chris as our relationship continues." On the other hand, if Joe has strong avoidance motives or if Joe and Chris had some false starts on their initial dates, he will be likely to adopt avoidance goals, such as, "I don't want us to be distant with each other or not become more intimate as our relationship continues." Or, in a discussion on the household budget, a husband who has strong approach social motives and recently has had several fun and loving interactions with his wife will be more likely to adopt approach goals, such as, "I want us to have an easy discussion and for both of us to be happy with the outcome." In contrast, a husband who has strong avoidance motives and has recently had a series of misunderstandings with his wife will be more likely to adopt avoidance goals, such as, "I want to avoid an argument and for neither of us to be dissatisfied with the outcome." Although the content of the goals in each example are very similar, the framing makes all the difference.

Our hierarchical model predicts that these approach and avoidance goals should be linked to different relationship outcomes. Approach relationship goals should be strongly associated with outcomes defined by the presence of social rewards, such as passion and intimacy. Thus, when individuals have strong approach goals in their relationships, positive interactions and relationship qualities are defined as those that provide such rewards as excitement, positive emotions, closeness, companionship, and understanding; negative interactions and relationships are those that fail to provide these rewards. Avoidance relationship goals should be strongly associated with outcomes defined by the presence of threats, such as criticism, rejection, and conflict. When individuals have strong avoidance goals in their relationship, positive interactions and relationships are defined as those that lack criticism, disagreement, rejection, and anxiety; negative interactions and relationships are those that possess these qualities. It is important to note that these two types of goals are independent of one another such that people may have strong approach and weak avoidance relationship goals, weak approach and strong avoidance relationships goals, strong approach and strong avoidance relationship goals, or weak approach and weak avoidance relationship goals. For example, if a person has strong approach and avoidance relationship goals, positive interactions and relationships are those that both provide rewards and lack of threats.

A major prediction of the model is that different processes should mediate the association between approach goals and outcomes and the association between avoidance goals and outcomes. Therefore, the processes that account for the link between approach goals and passion, for example, may not be the same processes that account for the link between avoidance goals and security. Finally, the model also outlines that the reward-based and threat-based relationship outcomes (i.e., intimacy, security) combine to form global assessments about relationship quality and form the basis for maintenance versus termination decisions. As the two long curved arrows in Figure 4.1 suggest, the model also predicts that approach goal strength should moderate how heavily these reward-based outcomes are weighted in global assessments and avoidance goal strength should moderate how heavily these threat-based outcomes are weighted in global assessments.

It is also important to note two of the model's features that are not easily represented in Figure 4.1. First, close relationships are dyadic and motivational accounts of relationship processes must go beyond the individual as the unit of analysis. In the context of a close relationship it is crucial to investigate the goals of not only one individual but also those of the partner and the unique interaction of the two. The second feature of a motivational model of close relationships is that close relationships exist, by definition, over time. As such, relationship goals are expected to fluctuate dynamically over time, responding to progress assessments and changes in the partner and the larger social environment. In the following sections I will review the evidence thus far supporting the links between approach and avoidance relationship goals, present results of studies investigating mediating processes, and discuss future directions and implications of the model.

Linking Approach and Avoidance Goals to Relationship Outcomes

The links between approach and avoidance goals and social and relationship outcomes have been investigated in several studies. Some of these studies have examined goals people have for their close relationships in general (e.g., "Deepening my relationships with my friends"; "Avoiding being hurt by my friends"), and other studies have focused on goals people have for a specific relationship, such as a romantic relationship (e.g., "Enhance the intimacy in my romantic relationship"). In a series of three studies, Gable (2006) examined several close relationships (family, friends, romantic partner). As predicted by the model in Figure 4.1, approach and avoidance motives and goals were associated with different outcomes.

Specifically, approach motives and goals were associated with positive affect toward close relationship partners and satisfaction with close relationships, concurrently and as they change over time. Avoidance motives and goals were associated with negative affect toward relationship partners and relationship insecurity, concurrently and as they change over time. Both approach and avoidance goals and motives predicted loneliness and negative and positive associations, respectively; loneliness is an outcome defined by a lack of incentives (meaningful ties) and threats (insecurity). It should be noted that we often find that avoidance goals are associated with outcomes that are negative in valence, even though the fulfillment of avoidance goals should lead to outcomes that are positive in valence, such as security and trust. As discussed in the subsequent section on mediating processes, our work has repeatedly shown that avoidance goals tend to lead to biases that ironically make threat outcomes (e.g., insecurity, rejection, loneliness) *more* likely.

In these studies it also important to note that the effects of relationship-specific motives and goals were significant predictors of outcomes even after controlling for general sensitivity to reward and punishment, such as individual differences in Behavioral Activation System (BAS) and Behavioral Inhibition System (BIS) sensitivities (Carver & White, 1994; Gray, 1990). In another set of studies Elliot, Gable, and Mapes (2006) focused on approach and avoidance goals for existing friendships. Consistent with Gable (2006), these studies found that strong approach friendship goals were positively associated with social satisfaction and negatively associated with loneliness. In addition, approach friendship goals were associated with positive changes in well-being over time, and avoidance friendship goals were associated with increased physical symptoms over time.

A series of studies have focused on romantic relationships and have shown that approach and avoidance goals for a specific relationship (or specific interactions) are also associated with distinct outcomes. In the first such study, Gable (2000) found that married couples who had strong avoidance goals also reported greater insecurity than those with weaker avoidance goals. In this same study, the strength of approach goals was also positively associated with more daily positive affect during marital interactions. Impett, Strachman, Finkel, and Gable (2008) also found that approach goals (but not avoidance goals) were strongly related to sexual desire in romantic relationships. Specifically, in three studies they found that adopting strong (relative to weak) approach goals in relationships predicted greater sexual desire over time. They showed that approach relationship goals buffered against declines in sexual desire over time and were associated with more sexual desire during sexual interactions. These associations were even stronger for women than men.

In a daily experience study, Gable and Poore (2008) found that individuals with strong approach goals were more satisfied on a daily basis (i.e., "Overall, how was your relationship today?" answered on a scale ranging from terrible to terrific) than those with weak approach goals. They also found that individuals with strong avoidance goals were less satisfied than those with weak avoidance goals on a daily basis. Again, the finding that overall satisfaction was predicted by both approach and avoidance goals parallels the loneliness findings previously reported such that global satisfaction is thought to be a combination of both the presence of rewards and the absence of threats. Impett and colleagues (in press) also examined the links between approach and avoidance goals in dating couples and daily relationship satisfaction and changes in global satisfaction over time. They found that individuals with stronger approach goals reported higher levels of daily relationship satisfaction and greater satisfaction over time than those with weaker approach goals for their relationships. Also, reflecting the dyadic nature of relationship goals, they found that the greatest satisfaction levels were found in couples in which both partners had strong approach goals. Similarly, individuals with stronger avoidance goals reported less satisfaction over time than those with weaker avoidance goals; those who had a partner who had strong avoidance goals were also less satisfied.

Studies focused on the goals that individuals have for specific interactions with their partners have also found that approach and avoidance motives predict outcomes in romantic relationships. Specifically, Impett, Gable, and Peplau (2005) examined dating couples' motives for enacting behaviors that are not preferred for the sake of their partners (i.e., everyday relationship sacrifices; Van Lange et al., 1997; enacting a behavior that is not preferred, such as accompanying a partner to a dull work function, not spending time with friends). They found that when people sacrificed for approach motives (e.g., to promote intimacy) they reported greater positive affect and relationship satisfaction. However, when they enacted the same behaviors for avoidance motives (e.g., to prevent my partner from becoming upset), they reported greater negative affect, lower relationship satisfaction, and more conflict. Sacrificing for avoidance motives seemed to erode relationships over time as well: Specifically, even when controlling for initial satisfaction, the more individuals sacrificed for avoidance motives over the course of the study, the less satisfied they were with their relationships at the 6-week follow-up assessment and the more likely they were to have broken up.

In summary, across several studies that employed diverse methods and assessed various close relationships such as friendships, romantic relationships, and family ties, approach and avoidance motives and goals have been associated with important relationship outcomes. Also, as predicted by the model, approach goals were more strongly associated with outcomes defined by the presence of incentives (e.g., positive affect, sexual desire) than avoidance goals. Avoidance goals were more strongly associated with outcomes defined by the absence of threats (e.g., negative affect, security) than approach goals. And approach and avoidance goals were associated with some outcomes that involve both incentives and threats, such as global satisfaction and loneliness. Finally, the results have been demonstrated in cross sectional, daily experience, and longitudinal data. In the next section, I review the processes hypothesized to link goals and motives to outcomes and review the evidence to date supporting these links.

Mediating Processes Linking Goals to Outcomes

Several studies have investigated the processes that link approach and avoidance relationship goals to outcomes. It is important to note that our model predicts that approach and avoidance relationship goals are relatively independent. The implication of this tenet is that the processes that link approach goals to outcomes are not necessarily the same processes that link avoidance goals to outcomes. For example, there is evidence that approach and avoidance goals are linked to variation in the experience of social events (e.g., Gable, Reis, & Elliot, 2000). Consistent with previous work on domain-general incentive and punishment motivations (individual differences in BAS and BIS), Gable (2006) found that approach social motives and goals were associated with increased exposure to social positive events. That is, strong approach relationship motives and goals predicted increased frequency of the occurrence of positive social events; approach relationship motives and goals did not predict the frequency of negative social events. The frequency of positive events mediated the link between approach relationship motives and outcomes.

On the other hand, avoidance relationship motives and goals were correlated with increased reactivity to negative social events (Elliot et al., 2006; Gable, 2006). That is, when negative social events did occur, those with strong avoidance relationship motives and goals rated them as more important and showed more changes in their well-being than those with weak avoidance goals. In short, a process of exposure to positive social events mediated between approach relationship goals and outcomes, whereas a process of reactivity to negative social events linked avoidance relationship goals to outcomes. Thus, there is tentative evidence that approach and avoidance motives are linked to social outcomes through different processes (see also Gable et al., 2000). Other recent research on social goals has examined how relationship goals predict outcomes through social cognitive and affective processes, such as biases in memory and interpretation of ambiguous information, the propensity for experiencing positive emotions, and the weighting of different types of information in global evaluations.

Memory and Interpretation of Ambiguous Information

A straightforward prediction of the relationship goals model is that social goals should influence the interpretation of ambiguous social information and later recall of that information. Specifically, approach goals may bias people to be more aware of and expectant of the potential incentives of their social environment and to recall this information more readily. Avoidance goals should bias people to be more aware of and expectant toward the potential threats in their social environment and to recall this information more readily. Memory of social information and biases in awareness are likely important processes in relationships. For example, Neuberg's (e.g., 1996) work has shown that expectancy–confirmation processes affect what information is sought from social partners, how people behave toward a social partner, and how the social partner behaves. In close relationships, people have repeated interactions with the same partner and thus have ample opportunity to form memories and expectancies regarding that partner. Thus, knowing how relationship goals may influence memories and the interpretation of new information seems particularly important.

Strachman and Gable (2006) investigated memory biases associated with relationship goals in two studies. In the first study, individual differences in the strength of goals for current friendships were assessed. Participants then read a story containing positive, negative, and neutral information regarding the interactions of two relationship partners. The results of Study 1 showed that people with strong avoidance social goals recalled more of the negative information in the story than those with weak avoidance goals. In Study 2, social goals for an upcoming interaction with a stranger were experimentally manipulated; half of the participants were given approach goals for a get-acquainted conversation, and half were given avoidance goals (e.g., try to make a good impression, try not to make a bad impression). They then were given a description ostensibly written by the other person. Those in the avoidance goal condition remembered more negative information about and expressed more dislike for the other person than those in the approach goal condition. These two studies provided good evidence that memory for social information is influenced by the focus of relationship goals (i.e., on incentives or threats). It is likely that this occurs in ongoing romantic relationships; however, empirical support for this is needed.

The social interactions and relationship events that people remember are subjective and often open to multiple interpretations in the first place.

For example, is a friend's quietness indicative of a pensive mood or lingering anger over a recent political disagreement? Strachman and Gable (2006) also analyzed participants' interpretations of positive, negative, and neutral information. Specifically, how people represented information from the story about two close relationship partners they had read about was examined. People with strong avoidance goals were more likely to interpret seemingly neutral and positive information from the story with a more negative spin than people with weak avoidance goals. For example, when describing a seemingly neutral part of the story, such as, "He picked her up at 10:00," those with high avoidance goals were likely to interpret that as being picked up late.

Affective Experience in Close Relationships

Previous research has found that approach motivation is closely tied to reports of positive affect such that those with strong approach motives and goals report high levels of positive affect on a daily basis and those with weak approach motives and goals report less positive affect on a daily basis. On the other hand, avoidance motivation is closely tied to reports of negative affect such that those with strong avoidance motives and goals report more negative affect on a daily basis than those with weak avoidance goals. More importantly, there is little evidence that approach goals predict negative affect and avoidance goals predict positive affect (Gable et al., 2000).[1]

In terms of ongoing relationships, Impett and colleagues (in press) conducted two studies investigating the role of positive emotions in the link between relationship goals and satisfaction in dating couples. In these studies, the strength of approach and avoidance goals dating partners had for the romantic relationship were assessed. In the first study they found that the strength of approach (but not avoidance goals) was positively associated with general positive affect as measured by the Positive and Negative Affect Schedule (PANAS; Watson, Clark, & Tellegen, 1988) on a daily basis. In the second study, they found that approach goals were also associated with relationship-specific positive emotions such as gratitude, love, and compassion. Moreover, they found that the strength of participants' approach goals was also positively associated with their partners' reports of positive emotions. In turn, the experience of positive emotions mediated the link between approach goals and relationship satisfaction. More research is needed to understand how avoidance goals

---

[1] There is recent evidence that approach motivation is associated with approach (versus withdrawal) emotions, most but not all of which are positive, and that avoidance motivation is associated with withdrawal emotions, not all of which are negative. For example, anger is an emotion related to approach motivation (for a review see Carver & Harmon-Jones, 2009), and relief is an emotion related to avoidance motivation (Carver, 2009).

and negative emotions might mediate the link between avoidance goals and relationship outcomes.

Weight of Social Information

Another important aspect of the relationship goals model depicted in Figure 4.1 is how incentive-based outcomes such as passion and threat-based outcomes such as security are combined in making more global evaluations in relationships. That is, it is likely that approach and avoidance relationships goals systematically influence the type of relationship information used in global evaluations. Specifically, those with strong approach goals should weigh the presence (or absence) of incentives in their relationships more heavily in global evaluations than those with weaker approach goals. And those with strong avoidance goals should place more weight on the presence (or absence) of threats in their relationships when making global evaluations than should those with weaker avoidance goals.

Gable and Poore (2008) conducted a signal-contingent daily experience study of people in dating relationships. Participants were beeped at several random intervals throughout the day and reported their feelings of passion for and security with their relationship partners at that very moment. At the end of the day they also reported their overall satisfaction with their relationships (on a scale from terrible to terrific). The results showed that individuals with strong approach goals weighed passion more heavily than those low in approach goals in their end-of-day reports of relationship satisfaction. Those with strong avoidance social goals weighed security more than those with weaker avoidance social goals in their end-of-day reports of satisfaction. Thus, on days people felt more passion than they typically felt, they reported increased relationship satisfaction only if they had strong approach relationship goals, and on days they felt more insecure than they typically felt, they reported less relationship satisfaction only if they had strong avoidance relationship goals. Consistent with previous work by Updegraff, Gable, and Taylor (2004) on general approach and avoidance goals and the weighting of positive and negative emotions, Gable and Poore found that the very definition of satisfaction (the presence of incentives or the absence of threats) was influenced by goals.

Summary and Future Directions

The aim of this chapter was to present a model of close relationship processes that addresses the regulation of both incentives and threats inherent in social bonds. The model is based on a solid foundation of evidence that the approach and avoidance distinction is fundamental to motivation. Moreover, the empirical literature has provided clear evidence for the existence of important incentives and threats in close relationships; thus, the domain of close relationships is particularly in need of a model that simultaneously addresses the regulation of both outcomes. As research on incentive and threat regulation in relationships continues, future investigations need to address three areas.

First, as previously noted, part of the definition of a close relationship is that it persists over time. Thus, examinations of how relationship partners evaluate progress on their approach and avoidance goals over time are likely to be particularly fruitful. As would be predicted by Carver and Scheier's (1982, 1990) control-process model, it is also very likely that a fundamental contributor to behavior and the cognitive and affective experience of people in close relationships will be their evaluation of goal progress. According to Carver and Scheier (1990), relationship partners likely compare their current level of goal attainment with a standard: If progress exceeds the standard, positive affect is experienced; if progress falls short of the standard, negative affect is experienced; and if progress is equal to the standard, no affect is experienced.[2] Moreover, the cues of progress and rate of progress are likely to be assessed differently for approach goals than for avoidance frameworks. For example, the husband who has the goal of not arguing with his wife is only one disagreement away from failure at any given time, regardless of how many fun and pleasant interactions he experiences with his wife. However, a husband who has an approach goal of spending quality one-on-one time with his wife grows closer to his goal with each date night scheduled.

A second area of focus of future research should be on the dyadic influence of partners' motives and goals. That is, how do one partner's approach and avoidance goals affect the other partner? Impett et al. (2005) reported data suggesting that this will be an important avenue of research. Specifically, when one partner perceived that the other was making everyday sacrifices for avoidance motives, he or she reported less well-being and lower relationship quality than when that partner perceived that the same sacrifices were made for approach reasons. In addition, Impett and colleagues (in press) showed that having a partner with strong avoidance relationship goals was associated with decreases in satisfaction over time. These data support the idea that the dyadic nature of close relationships dictates that both perceptions of the partner's motives and the partner's actual motives should influence relationship processes.

A final future direction in this line of research is the distinction between implicit and explicit goals. Up to this point, research on approach and avoidance goals in relationships has focused exclusively

[2] To understand the standard to which progress is compared, consideration of construct of comparison level outlined by interdependence theory (Kelley & Thibaut, 1978; Rusbult & Buunk, 1993) will likely be productive.

on explicit or consciously accessible goals. However, work from several other researchers on the power that seemingly nonconscious goal pursuit has on behavior in close relationships is compelling (e.g., Andersen, Reznik, & Manzella, 1996; Baldwin, Carrell, & Lopez, 1990; Fitzsimons & Bargh, 2003; Mikulincer, 1998; Scinta & Gable, 2007). For example, Fitzsimons and Fishbach (2010) recently reported that we show preferences and feel closer to relationship partners who are instrumental to our personal goal pursuit. Moreover, priming cues of progress on these goals leads to shifts in closeness to these partners. Future studies need to understand the implication of nonconscious motivation on approach and avoidance models.

Concluding Comments

Although the approach and avoidance motivation distinction has provided great insights in several domains in the psychological literature, it has rarely been considered in close relationships research. This is likely a lost opportunity. The framework explicitly describes the regulation of the inherent incentives and threats in close relationships and may offer insights into understanding different pathways to satisfaction and stability in close relationships. For example, some unstable and unsatisfying close relationships can be described as lacking incentives, and other unstable and unsatisfying close relationships can be described as teeming with threats. These two different states may be both a contributor to and result of the strength of approach and avoidance motives and goals. Moreover, different processes are likely operating to regulate incentives and threats, and any attempt to better these relationships needs to address both dimensions of the relationship.

Acknowledgment

Preparation of this chapter was facilitated by CAREER Grant #BCS 0444129 from the National Science Foundation.

References

Algoe, S., Gable, S. L., & Maisel, N. C. (2010). It's the little things: Everyday gratitude as a booster shot for romantic relationships. *Personal Relationships, 17,* 217–233.

Algoe, S. B., & Haidt, J. (2009). Witnessing excellence in action: The "other-praising" emotions of elevation, gratitude, and admiration. *Journal of Positive Psychology, 4,* 105–127.

Algoe, S. B., Haidt, J., & Gable, S. L. (2008). Beyond reciprocity: Gratitude and relationships in everyday life. *Emotion, 8*(3), 425–429.

Andersen, S. M., Reznik, I., & Manzella, L. M. (1996). Eliciting facial affect, motivation, and expectancies in transference: Significant-other representations in social relations. *Journal of Personality and Social Psychology, 71,* 1108–1129.

Aron, A., Norman, C. C., Aron, E. N., McKenna, C., & Heyman, R. E. (2000). Couples' shared participation in novel and arousing activities and experienced relationship quality. *Journal of Personality and Social Psychology, 78*(2), 273–284.

Atkinson, J. W. (Ed.) (1958). *Motives in fantasy, action, and society.* Princeton, NJ: Van Nostrand.

Atkinson, J. W., Heyns, R. W., & Veroff, J. (1954). The effect of experimental arousal of the affiliation motive on thematic apperception. *Journal of Abnormal & Social Psychology, 49*(3), 405–410.

Ayduk, O., Downey, G., Testa, A., Yen, Y., & Shoda, Y. (1999). Does rejection elicit hostility in rejection sensitive women? *Social Cognition Special Issue: Social Cognition and Relationships, 17*(2), 245–271.

Baldwin, M. W., Carrell, S. E., & Lopez, D. F. (1990). Priming relationship schemas: My advisor and the Pope are watching me from the back of my mind. *Journal of Experimental Social Psychology, 26,* 435–454.

Baron, K. G., Smith, T. W., Butner, J., Nealey-Moore, J., Hawkins, M. W., & Uchino, B. N. (2007). Hostility, anger, and marital adjustment: Concurrent and prospective associations with psychosocial vulnerability. *Journal of Behavioral Medicine, 30*(1), 1–10.

Bartels, A., & Zeki, S. (2000). The neural basis of romantic love. *Neuroreport, 11*(17), 3829–3834.

Bartels, A., & Zeki, S. (2004). The neural correlates of maternal and romantic love. *Neuroimage, 21*(3), 1155–1166.

Baumeister, R. F., & Leary, M. R. (1995). The need to belong: Desire for interpersonal attachments as a fundamental human motivation. *Psychological Bulletin, 117,* 497–529.

Bowlby, J. (1969). *Attachment and loss: Vol 1. Attachment.* New York: Basic Books.

Boyatzis, R. E. (1973). Affiliation motivation. In D. C. McClelland & R. S. Steele, (Eds.) *Human motivation: A book of readings* (pp. 252–276). Morristown, NJ: General Learning Press.

Cacioppo, J. T., & Patrick, W. (2008). *Loneliness: Human nature and the need for social connection.* New York: W. W. Norton & Co.

Carter, C. S. (1998). Neuroendocrine perspectives on social attachment and love. *Psychoneuroendocrinology, 23*(8), 779–818.

Carver, C. S. (2009). Threat sensitivity, incentive sensitivity, and the experience of relief. *Journal of Personality, 77*(1), 125–138.

Carver, C. S., & Harmon-Jones, E. (2009). Anger is an approach-related affect: Evidence and implications. *Psychological Bulletin, 135*(2), 183–204.

Carver, C. S., & Scheier, M. F. (1982). Control theory: A useful conceptual framework for personality-social, clinical, and health psychology. *Psychological Bulletin, 92,* 111–135.

Carver, C. S., & Scheier, M. F. (1990). Origins and functions of positive and negative affect: A control-process view. *Psychological Review, 97,* 19–35.

Carver, C. S., & White, T. L. (1994). Behavioral inhibition, behavioral activation, and affective responses to impending reward and punishment: The BIS/BAS scales. *Journal of Personality and Social Psychology, 67,* 319–333.

Collins, N. L., & Feeney, B. C. (2000). A safe haven: An attachment theory perspective on support seeking and caregiving in adult romantic relationships. *Journal of Personality and Social Psychology, 58,* 644–663.

Danner, D. D., Snowdon, D. A., & Friesen, W. V. (2001). Positive emotions in early life and longevity: Findings from the nun study. *Journal of Personality and Social Psychology, 80*(5), 804–813.

Davis, D., Shaver, P. R., & Vernon, M. L. (2004). Attachment style and subjective motivations for sex. *Personality and Social Psychology Bulletin, 30,* 1076–1090.

Dickerson, S. S., Gable, S. L., Irwin, M. R., Aziz, N., & Kemeny, M. E. (2009). Social-evaluative threat and proinflammatory cytokine regulation: An experimental laboratory investigation. *Psychological Science, 20,* 1237–1244.

Dickerson, S. S., & Kemeny, M. E. (2004). Acute stressors and cortisol responses: A theoretical integration and synthesis of laboratory research. *Psychological Bulletin, 130*(3), 355–391.

Downey, G., Feldman, S., & Ayduk, O. (2000). Rejection sensitivity and male violence in romantic relationships. *Personal Relationships, 7*(1), 45–61.

Downey, G., Freitas, B. L., Michaelis, B., & Khouri, H. (1998). The self-fulfilling prophecy in close relationships: Rejection sensitivity and rejection by romantic partners. *Journal of Personality and Social Psychology, 75,* 545–560.

Eisenberger, N. I., Jarcho, J. M., Lieberman, M. D., & Naliboff, B. D. (2006). An experimental study of shared sensitivity to physical pain and social rejection. *Pain, 126,* 132–138.

Eisenberger, N. I., Lieberman, M. D., & Williams, K. D. (2003). Does rejection hurt? An fMRI study of social exclusion. *Science, 302,* 290–292.

Elliot, A. J. (2006). The hierarchical model of approach-avoidance motivation. *Motivation and Emotion, 30,* 111–116.

Elliot, A. J., Gable, S. L., & Mapes, R. R. (2006). Approach and avoidance motivation in the social domain. *Personality and Social Psychology Bulletin, 32,* 378–391.

Feeney, B. C., Cassidey, J., & Ramos-Marcuse, F. (2008). The generalizaiton of attachment represenations to new social situations: Predicting behavior during initial interactions with strangers. *Journal of Personality and Social Psychology, 95*(6), 1481–1498.

Fitzsimons, G. M., & Bargh, J. A. (2003). Thinking of you: Nonconscious pursuit of interpersonal goals associated with relationship partners. *Journal of Personality and Social Psychology, 84,* 148–164.

Fitzsimons, G. M., & Fishbach, A. (2010). Shifting closeness: Interpersonal effects of personal goal progress. *Journal of Personality and Social Psychology, 98*(4), 535–549.

Gable, S. L. (2000). *Appetitive and aversive social motivation.* Unpublished doctoral dissertation, University of Rochester.

Gable, S. L. (2006). Approach and avoidance social motives and goals. *Journal of Personality, 74,* 175–222.

Gable, S. L., & Berkman, E. T. (2008). Making connections and avoiding loneliness: Approach and avoidance social motives and goals. In A. J. Elliot (Ed.), *Handbook of approach and avoidance motivation* (pp. 203–216). New York: Psychology Press.

Gable, S. L., & Poore, J. (2008). Which thoughts count? Algorithms for evaluating satisfaction in relationships. *Psychological Science, 19,* 1030–1036.

Gable, S. L., & Reis, H. T. (2001). Appetitive and aversive social interaction. In J. H. Harvey & A. E. Wenzel (Eds.), *Close romantic relationships: Maintenance and enhancement* (pp. 169–194). Mahwah, NJ: Erlbaum.

Gable, S. L., Reis, H. T., & Elliot, A. J. (2000). Behavioral activation and inhibition in everyday life. *Journal of Personality and Social Psychology, 78,* 1135–1149.

Gillath, O., Bunge, S. A., Shaver, P. R., Wendelken, C., & Mikulincer, M. (2005). Attachment-style differences in the ability to suppress negative thoughts: Exploring the neural correlates. *Neuroimage, 28*(4), 835–847.

Gottman, J. M. (1994). *What predicts divorce?* Hillsdale, NJ: Erlbaum.

Gray, J. A. (1990). Brain systems that mediate both emotion and cognition. *Cognition and Emotion, 4,* 269–288.

Harker, L., & Keltner, D. (2001). Expressions of positive emotion in women's college yearbook pictures and their relationship to personality and life outcomes across adulthood. *Journal of Personality and Social Psychology, 80*(1), 112–124.

House, J. S., Landis, K. R., & Umberson, D. (1988). Social relationships and health. *Science, 241,* 540–545.

Huston, T. L., Caughlin, J. P., Houts, R. M., Smith, S. E., & George, L. J. (2001). The connubial crucible: Newlywed years as predictors of marital delight, distress, and divorce. *Journal of Personality and Social Psychology, 80,* 237–252.

Impett, E., Gable, S. L., & Peplau, L. A. (2005). Giving up and giving in: The costs and benefits of daily sacrifice in intimate relationships. *Journal of Personality and Social Psychology, 89,* 327–344.

Impett, E. A., Gordon, A. M., Kogan, A., Oveis, C., Gable, S. L., & Keltner, D. (2010). Moving toward more perfect unions: Daily and long-term consequences of approach and avoidance goals in romantic relationships. *Journal of Personality and Social Psychology, 99*(6), 948–963.

Impett, E., Strachman, A., Finkel, E., & Gable, S. L. (2008). Maintaining sexual desire in intimate relationships: The importance of approach goals. *Journal of Personality and Social Psychology, 94,* 808–823.

Insel, T. R. (2000). Toward a neurobiology of attachment. *Review of General Psychology, 4,* 176–185.

Kelley, H. H., & Thibaut, J. W. (1978). *Interpersonal relations: A theory of interdependence.* New York: Wiley.

Kiecolt-Glaser, J. K. (1999). Stress, personal relationships, and immune function: Health implications. *Brain, Behavior, and Immunity, 13*(1), 61–72.

Klinger, E. (1977). *Meaning & void: Inner experience and the incentives in people's lives.* Minneapolis: University of Minnesota Press.

Lyubomirsky, S., King, L., & Diener, E. (2005). The benefits of frequent positive affect: Does happiness lead to success? *Psychological Bulletin, 131*(6), 803–855.

McAdams, D. P., Healy, S., & Krause, S. (1984). Social motives and patterns of friendship. *Journal of Personality and Social Psychology, 47,* 828–838.

Mehrabian, A. (1976). Questionnaire measures of affiliative tendency and sensitivity to rejection. *Psychological Reports, 38,* 199–209.

Mikulincer, M. (1998). Attachment working models and the sense of trust: An exploration of interaction goals and affect regulation. *Journal of Personality and Social Psychology, 74,* 1209–1224.

Miller, N. E. (1959). Liberalization of basic S-R concepts: Extensions to conflict behavior, motivation, and social learning. In S. Koch (Ed.), *Psychology: A study of a science, Study 1* (pp. 198–292). New York: McGraw-Hill.

Neuberg, S. L. (1996). Expectancy influences in social interaction: The moderating role of social goals. In P. M. Gollwitzer & J. A. Bargh (Eds.), *The psychology of action: Linking cognition and motivation to behavior* (pp. 529–552). New York: Guilford Press.

Notarius, C., & Markman, H. (1993). *We can work it out: Making sense of marital conflict.* New York: G. P. Putnam's Sons.

Panksepp, J. (1998). *Affective neuroscience: The foundations of human and animal emotions.* London: Oxford University Press.

Pinsker, H., Nepps, P., Redfield, J., & Winston, A. (1985). Applicants for short-term dynamic psychotherapy. In A. Winston (Ed.), *Clinical and research issues in short-term dynamic psychotherapy* (pp. 104–116). Washington, DC: American Psychiatric Association.

Reis, H. T., Collins, W. A., & Berscheid, E. (2000). The relationship context of human behavior and development. *Psychological Bulletin, 126,* 844–872.

Reis, H. T., & Gable, S. L. (2003). Toward a positive psychology of relationships. In C. L. Keyes & J. Haidt (Eds.), *Flourishing: Positive psychology and the life well-lived* (pp. 129–159). Washington, DC: American Psychological Association.

Rusbult, C. E., & Buunk, B. P. (1993). Commitment processes in close relationships: An interdependence analysis. *Journal of Social and Personal Relationships. Special Issue: Relational Maintenance, 10*(2), 175–204.

Schneirla, T. C. (1959). An evolutionary and developmental theory of biphasic processes underlying approach and withdrawal. *Nebraska Symposium on Motivation* (Vol. 7, pp. 1–43). Lincoln: University of Nebraska Press.

Scinta, A., & Gable, S. L. (2007). Automatic and self-reported attitudes in romantic relationships. *Personality and Social Psychology Bulletin, 33*(7), 1008–1022.

Strachman, A., & Gable, S. L. (2006). What you want (and don't want) affects what you see (and don't see): Avoidance social goals and social events. *Personality and Social Psychology Bulletin, 32,* 1446–1458.

Taylor, S. E, Klein, L. C., Lewis, B. P., Gruenewald, T. L., Gurung, R. A. R., & Updegraff, J. A. (2000). Biobehavioral responses to stress in females: Tend-and-befriend, not fight-or-flight. *Psychological Review, 107,* 411–429.

Thibaut, J. W., & Kelley, H. H. (1959). *The social psychology of groups.* New York: Wiley.

Tran, S., & Simpson, J.A. (2009). Pro-relationship maintenance behaviors: The joint roles of attachment and commitment. *Journal of Personality and Social Psychology, 97*(4), 685–698.

Tsapelas, I., Aron, A., & Orbuch, T. (2009). Marital boredom now predicts less satisfaction 9 years later. *Psychological Science, 20*(5), 543–545.

Uchino, B. N., Holt-Lunstad, J., Uno, D., & Flinders, J. B. (2001). Heterogeneity in the social networks of young and older adults: Prediction of mental health and cardiovascular reactivity during acute stress. *Journal of Behavioral Medicine, 24*(4), 361–382.

Updegraff, J. A., Gable, S. L., & Taylor, S. E. (2004). What makes experiences satisfying? The interaction of approach-avoidance motivations and emotions in well-being. *Journal of Personality and Social Psychology, 86*, 496–504.

Van Lange, P. A. M., Rusbult, C. E., Drigotas, S. M., Arriaga, X. M., Witcher, B. S., & Cox, C. L. (1997). Willingness to sacrifice in close relationships. *Journal of Personality and Social Psychology*, 72, 1373–1395.

Watson, D., Clark, L. A., & Tellegen, A. (1988). Development and validation of brief measures of positive and negative affect: The PANAS scales. *Journal of Personality and Social Psychology, 54*, 1063–1070.

Zak, P. J., Kurzban, R., & Matzner, W. T. (2005). Oxytocin is associated with human trustworthiness. *Hormones and Behavior, 48*(5), 522–527.

CHAPTER 5

Social Baseline Theory and the Social Regulation of Emotion

LANE BECKES

University of Virginia

JAMES A. COAN

University of Virginia

Social relationships are powerfully associated with human health and well-being. Those who are socially isolated suffer a host of difficulties, from emotional pain to increased risk for illness and even death (House, Landis, & Umberson, 1988). By contrast, those who are embedded in a rich social network—and particularly those who inhabit satisfying close relationships—enjoy attenuated stress-related autonomic and hypothalamic-pituitary-adrenal (HPA) axis activity (Eisenberger, Taylor, Gable, Hilmert, & Lieberman, 2007; Flinn & England, 1997; Lewis & Ramsay, 1999) and lower risk for both physical and psychological maladies (Moak & Agrawal, 2010). Although many things may contribute to the salubrious effects of social relationships, there is increasing awareness that a major factor is the provision of perceived security that regulates negative emotion and physiological reactivity (Coan, 2008). We think of this as the social regulation of emotion and believe further that the emotion-regulatory effects of social relationships represent a major human adaptation—one that forms part of the foundation of the human ecology, which can itself be found, more than any specific terrestrial environment or diet, in the company of other humans.

The Self-Regulation of Emotion Is Powerful and Costly

The ability to regulate our own emotions has provided humans with a distinct advantage over other animals. Even in a context of emotional intensity, humans are capable of soothing themselves for a variety of purposes, including patiently awaiting temporally distant rewards, conserving effort, and deceiving others (Gross & Thompson, 2007). Moreover, individual differences in emotion-regulation capabilities are consequential. In children, the ability to regulate emotion is associated with fewer externalizing problems such as aggression, conduct disorder, and oppositional defiant disorder (e.g., Crowe & Blair, 2008; Beauchaine, Gatzke-Kopp, & Mead, 2007; Hill, Degnan, Calkins, & Keane, 2006; Rydell, Berlin, & Bohlin, 2003) as well as decreased risk for affective disorders (Buckner, Mezzacappa, & Beardslee, 2009; Dennis, Brotman, Huang, & Gouley, 2007). Effective emotional self-regulation is no less beneficial to adults, where it is similarly associated with enhanced life satisfaction, decreased affective psychopathology, and better overall health (Haga, Kraft, & Corby, 2009; Smyth & Arigo, 2009). Many psychotherapeutic interventions emphasize self-regulation. In particular, cognitive behavioral therapy and mindfulness meditation help individuals develop their capacity to regulate their own emotional responses (Lykins & Baer, 2009; Smyth & Arigo, 2009; Suveg, Sood, Comer, & Kendall, 2009).

Indeed, human self-regulation capabilities are so powerful that we find it is useful to think of humans as the cheetahs of emotional self-regulation. Just as the cheetah is Earth's fastest land animal (capable of reaching speeds of up to 75 miles per hour), no other species on the planet is capable of crafting a regulatory cognition that even approximates the phrase, "It's only a movie." Less appreciated, however, is that the cheetah analogy extends to another important aspect of self-regulation. Just as the cheetah can sustain its top speed only for short bursts of time, human self-regulation abilities are difficult to sustain for long periods (Gailliot & Baumeister, 2007). Several experiments by Baumeister and colleagues suggest that self-regulation depletes some kind of computational or physiological resource. In these experiments, the deployment of self-control in one situation appears to decrease self-regulation capabilities in another. For example, when compared with simply solving math problems, engaging in a thought-suppression task makes people more likely to drink a free alcoholic beverage before a driving evaluation (Muraven, Collins, & Nienhaus, 2002). Some have argued that this limitation is a function of blood glucose concentration, which is thought to diminish as a function of neural—particularly prefrontal—activity (Gailliot & Baumeister, 2007). This argument has more recently been challenged in two important ways. First, blood glucose levels in the brain probably do not change enough during self-regulation tasks to account for the apparent depletion effects that many have observed (Kurzban, 2010). Second, simple experimental manipulations—such as presenting subjects with a gift (Tice, Baumeister, Shmueli, & Muraven, 2007) or simply persuading subjects that willpower is an unlimited resource (Job, Dweck, & Walton, 2010)—can reduce or eliminate these depletion effects.

All of this suggests to us that (1) self-regulation is indeed costly (or depletion effects would not be so commonly observed); (2) the cost of self-regulation is unlikely to be a function of a specifically proximal resource limitation (meaning that apparent depletion effects can in essence be overridden when needed); (3) in any case, because self-regulation is apparently costly in some important if poorly understood way, the brain is designed to avoid engaging in it whenever possible; and (4) the brain probably reflexively uses a variety of heuristic cues to decide when it can conserve its regulatory capabilities instead of deploying them. In this way, rather than being beholden to some specific quantity of a metabolic resource (like glucose), the brain is probably designed to update its "budget" regarding self-regulatory resources, with an eye toward conservation, whenever possible. We will argue in this chapter that one of the human brain's primary sources of information for economizing its cognitive and regulatory activity is the degree of proximity to social resources (Coan, 2008, 2010). We will frame our argument in terms of our own empirical work and a conceptual frame borrowed from behavioral ecology and the study of perception–action links.

Social Regulation of the Neural Response to Threat

Many mammals regulate emotion through social contact and proximity (Fogel, 1993). Social contact exerts a significant impact on health and well-being, lowering resting blood pressure (Uchino, Holt-Lunstad, Uno, & Betancourt, 1999), decreasing risk of carotid artery atherosclerosis (Knox et al., 2000), minimizing salivary cortisol responses (Turner-Cobb, Sephton, Koopman, Blake-Mortimer, & Spiegel, 2000), and enhancing immune functioning (Lutgendorf et al., 2005; see Uchino, 2006, for a review). Higher levels of social integration even reduce age-adjusted risk of death (House et al., 1988).

Our lab has begun to systematically explore the social regulation of emotion using functional magnetic resonance imaging (fMRI), which measures changes in blood flow within the brain by contrasting blood deoxyhemoglobin levels across two or more experimental situations, a process referred to as blood-oxygen-level dependence (BOLD). In the first of these studies (Coan, Schaefer, & Davidson, 2006), functional brain images were collected from 16 married women, selected for very high marital satisfaction, who we placed under the threat of mild shock during each of three conditions: (1) holding the hand of their relational partner, (2) holding the hand of a stranger, or (3) lying alone in the scanner. Relationship satisfaction was measured with the Dyadic Adjustment Scale (DAS). Self-reports indicated that unpleasantness was lowest in the partner hand holding condition, relative to either stranger hand holding or no hand holding. Interestingly, both stranger and relational partner hand holding reduced subjective arousal relative to the alone condition.

Subsequent analysis of threat-related brain activity suggested that the brain was highly active when threats were faced alone, significantly less active during either stranger or partner hand holding, and less active still during partner hand holding. Specifically, regions of the brain that are likely involved in the modulation of arousal and bodily preparation for action, such as the ventral anterior cingulate cortex (vACC), posterior cingulate cortex, postcentral gyrus, and supramarginal gyrus, were all less responsive to threat cues during hand holding, regardless of whose hand was being held. In addition to these effects, however, partner hand holding also attenuated threat responding in the dorsolateral prefrontal cortex, superior colliculus, caudate, and nucleus accumbens—all regions associated with threat vigilance and self-regulation.

This pattern of findings closely mirrored those of participant self-reports, which suggested that physiological arousal was sensitive to any hand holding but that emotional valence was sensitive only to partner hand holding. Given that the sample consisted only of highly satisfied couples, it was surprising that relationship quality (DAS) was negatively correlated with threat responding in brain regions critical to the status and regulation of the body in response to stress, such as the right anterior insula ($r = -.47$), the left superior frontal gyrus ($r = -.59$), and the hypothalamus ($r = -.46$). More surprising still was that these negative correlations were observed only during the partner hand holding condition. Taken together, these findings provide strong evidence that most threat-responsive brain areas are less active when experiencing physical contact with another person. Moreover, the effect is larger and more widespread when that other person is a relational partner, and it is yet larger among individuals in the highest-quality relationships.

The Down-Regulation Model

The down-regulation model is in many ways the implicitly assumed model of emotion regulation, including social emotion regulation. This model postulates that some additional regulatory circuit inhibits circuits that are automatically threat responsive. Applied to the hand holding study previously discussed, this would suggest that the threat cue activated a widespread threat system and that hand holding activated an additional system that exerted a down-regulatory influence on the threat system that was already activated. For this model to be correct, of course, there must be the mediating circuit that is both activated by hand holding per se and is capable of down-regulating a variety of neural regions involved in the response to threat. Indeed, additional down-regulatory circuits may become active as a function of increasing familiarity with the provider of social support.

If this were true, it would seem likely that our brain would be most active (from a regulatory perspective) when receiving hand holding from a close relational partner with whom we share a very high-quality relationship. This is because we would have both excitatory and inhibitory activity occurring at once for a variety of threat-responsive regions, an effect that would resemble pressing on the accelerator of a car while simultaneously pressing on the brake. Although this is what we would predict, this is not what we would find. In fact, in the hand holding study just described, no neural circuits were found that were, independent of the presence of a threat cue, simply more active during hand holding, during hand holding associated with a relational partner, or during hand holding as a function of relationship quality. The brain was simply more active when facing threat alone relative to when facing threat coupled with social support. Thus, the down-regulation model of social support was not supported by the data.

The Social Baseline Model

The social baseline model questions the assumptions inherent in the down-regulation model by suggesting that the brain *assumes* proximity to social networks and relational partners. If this is the case, social proximity—not the alone condition—is the baseline, and social support acts not by exerting a regulatory force so much as by returning an organism to its baseline state. From this perspective, it is being alone that constitutes the special case, not the social support condition, that is in fact normative. The social baseline model requires a seemingly subtle but actually powerful change of perspective, much like a figure–ground illusion. It assumes that being alone increases threat sensitivity as opposed to assuming that being with others decreases threat sensitivity—two perspectives that are no more identical than the famous Rubin vase and the two faces that frame it. We expect that more is going on in a participant's environment when they are with a close friend during a threatening situation, because there are literally more perceivable stimuli, but it may be that from the brain's perspective there is actually *less* going on.

This perspective appeals to the principle of *economy in action* (cf. Proffitt, 2006), which states that organisms must consume more energy than they use to survive—an imperative that leads to the conservation of resources whenever possible (cf. Krebs & Davies, 1993; Proffitt, 2006). Proffitt and colleagues have argued that perception–action links are often tied to the economy of action, demonstrating, for example, that individuals perceive hills to be steeper and distances farther away if they are wearing a heavy backpack. According to Proffitt, the bodily perception of the heavy backpack translates to increased perceived physiological load, which causes a perceptual shift in the geographical features with which the individual must cope. All of this perceptual information is used to update the cost of engaging in the corresponding potential actions—climbing a steep hill or walking a long distance. Thus, with increased weight to carry, the hill appears steeper, and more motivation must be marshaled if the hill is going to be climbed. Put another way, without the backpack, a simple curiosity in what the top of the hill looks like may be sufficient to motivate climbing it. With the backpack on, however, simple curiosity might not provide sufficient motivation—a payoff commensurate with the additional cost is required. Because energy management is a critical aspect of daily living, and a major pressure in evolution, organisms have evolved to calculate (in mostly implicit ways) the perceived cost–benefit ratio of any given action or investment of resources, including, we believe, the self-regulation of emotion.

Social baseline theory (SBT; Coan, 2010) suggests that for humans being alone is like carrying a heavy backpack. The fundamental premise of SBT is that social proximity and interaction constitute the baseline human ecology and that socially mediated forms of emotion regulation are sufficiently powerful, widespread across a number of animal species, economical, and unconditioned to be considered the default human emotion-regulation strategy. From the perspective of SBT, just as salamanders are born with physiological (and obviously implicit) expectations of finding moist, cool, dark spaces to inhabit, humans are born with physiological, behavioral, and psychological expectations of human contact through touch and expression—of individuals with whom to share resources, goals, attention, and regulation (Kudo & Dunbar, 2001; Sbarra & Hazan, 2008; Tomasello, 2009). Indeed, social proximity and interaction are unconditionally reinforcing to humans in much the same way as water, food, and oxygen (Coan, 2008). It follows, then, that danger and difficulty increase as a function of distance from social support.

We use social cues to guide us in making decisions about the economy of certain actions, which in turn guides the activation of neural circuits commensurate with carrying out those actions. Because social proximity is the baseline situation, less effort is needed in terms of vigilance to threat and emotion regulation when we are with trusted others than when we are alone. When it comes to emotion regulation in particular, our efforts are normatively contracted out to close others who effectively act as surrogate prefrontal cortices. This allows us to achieve our regulatory goals at a limited cost. SBT refers to this process as *load sharing* and argues that it is largely a function of familiarity, interdependence, and interpersonal conditioning. With a moment's reflection, it is very easy to illustrate and understand. If an individual is confronted with four problems in his immediate environment, for example, then he must solve all of those problems himself if no one is there to help him. If a stranger is present, it may be that at least some of the load—say, a single one of those problems—can be "contracted out" to the stranger, leaving only three problems. If the social resource is familiar and predictable, the number of problems may reduce to two, and if the person is someone with whom a high-quality relationship is shared then only a single problem of the possible four may require solving entirely independently.

Examples of social emotion regulation are most obvious in infancy, where the regulation of fundamental physiological needs is achieved through caregiver responses to the infant's negative affect. As the infant develops through toddlerhood and beyond, the regulation of physiological needs via the child's affect gradually evolves into the regulation of the child's affect per se (Hofer, 2006). This occurs in tandem with the infant's neural development, which, through these years, is characterized by rapid expansion and tuning—a putatively (though not indisputably) critical period where the child is beginning to form expectations and implicit beliefs about the environment he or she can expect to face while developing toward independence.

Importantly, one of the least developed regions of the brain immediately following birth is the prefrontal cortex—the region of the brain most frequently and powerfully associated with self-regulation, including the self-regulation of emotion (Ochsner, Bunge, Gross, & Gabrieli, 2002; Ochsner & Gross, 2005). Humans are utterly dependent upon adult caregivers at birth, but even as they develop the means to be relatively independent physiologically and behaviorally they remain dependent on adults for many years because of underdeveloped reasoning and regulatory abilities yoked to similarly underdeveloped prefrontal cortices. A child of 7 years is physically capable of navigating a complex urban environment or a cross-country trip on public transportation, but apart from the base of knowledge required for such tasks (which is in fact minimal) she would not be expected to be particularly good at regulating her anxiety or at exercising rational judgments about how best to respond to unexpected dangers. Adult caregivers assist with these needs by loaning their children prefrontal effort. If children are incapable of regulating themselves at a frightening movie, parents can hold their hands and do the regulatory work for them by reminding them, for example, that the action in the movie isn't actually real and that the children are actually safe because the parents will protect them. At both the experiential and neural levels, this is effortful and (as many parents well know) even potentially exhausting.

Mediating Mechanisms?

As reviewed already, we do not think the down-regulation model is capable of explaining the results of the previously discussed hand holding study because those data did not reveal any neural activations positively correlated with hand holding per se, through which social regulation effects were mediated. Nevertheless, there must be some mechanism capable of identifying the presence of conspecifics, particularly relational partners, as well as a mechanism linking the general perception of plentiful social resources to attenuated threat reactivity.

One potential mechanism mediating the decreased responsiveness to threat during social contact is the neuropeptide oxytocin (OT). In a variety of species, including humans, OT plays a central role in social behavior. For example, OT is often observed to be released during pleasurable social contact, may be necessary for establishing and maintaining social bonds, particularly among monogamous species, and appears to be sufficient in many cases for increasing feelings of trust and inhibiting feelings of fear (Insel & Fernald, 2004; Taylor, 2006). These findings make it a natural candidate mechanism for the social regulation of neural threat responding. In fact, Kirsch and colleagues (Kirsch et al., 2005) found direct evidence for OT's role in the regulation of threat responses. Half their participants were given a placebo, and the other half were given OT via a nasal spray. The participants who received OT had significantly less BOLD response in the amygdala to negative emotional pictures than those in the placebo condition, indicating a reduced threat response as a function of OT administration. This indicates that OT may reduce threat vigilance. Given OT's tendency to release in the presence of social stimuli, social contact may reduce an individual's need to self-regulate in a manner consistent with SBT. If this is correct, then people should rely less on their prefrontal cortex to regulate their emotion when social resources are high.

Other work suggests a role for endogenous opioids, particularly in the dorsal ACC (dACC). For example, Eisenberger and colleagues (2007) observed that the dACC is sensitive to the availability of social resources, with important implications for how individuals respond to threatening stimuli. Specifically, Eisenberger et al. reported that greater daily levels of perceived social support were associated with lower levels of threat-related activity in the dACC. These authors point out that the dACC may be desensitized by social experience through repeated exposure to endogenous opioids. Indeed, the dACC has a high density of opioid receptors, and enogenous opioids are unconditionally released in response to positive social experiences (Panksepp, 1998; Panksepp, Nelson, & Siviy, 1994). Taken together, oxytocinergic activity in regions such as the hypothalamus, nucleus accumbens, and amygdala as well as endogenous opioid activity in the dACC are exciting possibilities. Still, a detailed delineation of the neural mechanisms of social emotion regulation awaits a great deal of additional research.

Summary and Conclusions

Human beings are powerful self-regulators, but self-regulation via the prefrontal cortex is exhausting, is probably costly in ways that are only recently well understood, and is likely ineffective as a sustained regulatory strategy. Alternatively, social contact and proximity appears to curb the need for self-regulation by reducing the need for negative affect and threat responding. We have suggested that this pattern points to social proximity as a likely baseline emotion-regulatory strategy for humans.

This line of thinking presents many opportunities for future research and theory development. Attachment theory in particular provides fertile ground for the integration of SBT ideas and hypotheses. For example, insecurely attached individuals should have different perceptions of the predictability of others as social resources relative to secure individuals. Under threat conditions, then, we should expect that the neural response to threat during hand holding is highly sensitive to attachment style, especially in intact relationships. A critical prediction of SBT is the conservation of computational or metabolic resources devoted to self-regulation, and although our initial hand holding fMRI provides some glimpses of such conservation, future work should seek to test this position in other ways. For example, according to Baumeister and colleagues, we might expect smaller changes in circulating blood glucose following a stressful self-regulation task in the presence of active social support. Or failing a direct impact on circulating blood glucose (an uncertain and in any case disputed potential proxy measure of cognitive effort), might stress lead to greater consumption of resources (concretely, eating more food), reflecting a change in resource "budgeting" as a result of increased self-regulatory demand? And might social emotion regulation mitigate this increase in consumption? Alternatively, self-regulation likely exerts a negative impact on other concomitant cognitive activities—might social forms of emotion regulation mitigate this impact as well? Future investigations should also explore infant emotion-regulation and neural development. Given the relative lack of development of the prefrontal cortex in the infant brain, social proximity is a fundamental strategy infants use to regulate their own affect. More needs to be known about how such social affect regulation impacts neural development and predicts the neural response to threat throughout life and how that development relates to developmental trajectories of attachment bonding.

Applied questions and interventions may also be informed by this line of inquiry. For example, might it be possible to develop a neural assay of social support by imaging the social regulation of hypothalamic activity during stress? Such information may lead to significant progress in predicting the effect of a person's social support network on a variety of health outcomes including response to medical treatment, risk of depression, and physical resilience to disease. Having strong social supports may prove critical in determining people's abilities to manage pain related to arthritis, cancer treatment, and a variety of other health problems. Various interventions could be used to improve the social regulation of emotion such as interdependence training for couples or possibly social capital development for communities. More can be done to understand how social resources can be mobilized to reduce the stress and health impact of major life transitions such as going to college, becoming a parent, or beginning retirement.

Many real-world anecdotes point to the potential for social forms of emotion regulation. Indeed, anecdotal reports one of us (JAC) collected during the initial hand holding study suggest that physical touch may play a more significant role in an individual's psychological state than expected. For example, one of the original hand holding study participants left the scanner crying and, when asked what was wrong, reported that the combination of threat and soothing from her husband caused her to remember the way her husband held her hand during labor—a memory that brought her tears of joy. In another example, an individual who hadn't participated in the hand holding study but had read about it in the popular press sent a letter to the laboratory describing her experience of coping with her husband's cancer. In it, she noted that "he never holds my hand; it is not like him. But after this surgery and all the time in the hospital, he constantly wants me to hold his hand. He reaches for me all the time." As these anecdotes indicate, there is something powerful about the physical touch of another human being, especially when we are in a time of need. We think that something is the social regulation of emotion, and we look forward to understanding it in far greater detail.

References

Beauchaine, T. P., Gatzke-Kopp, L., & Mead, H. K. (2007). Polyvagal theory and developmental psychopathology: Emotion dysregulation and conduct problems from preschool to adolescence. *Biological Psychology*, 74, 174–184.

Buckner, J. C., Mezzacappa, E., & Beardslee, W. R. (2009). Self-regulation and its relations to adaptive functioning in low-income youths. *American Journal of Orthopsychiatry, 79*, 19–30.

Coan, J. A. (2008). Toward a neuroscience of attachment. In J. Cassidy & P. R. Shaver (Eds.), *Handbook of attachment: Theory, research, and clinical applications* (2nd ed., pp. 241–265). New York: Guilford Press.

Coan, J. A. (2010). Adult attachment and the brain. *Journal of Social and Personal Relationships, 27*, 210–217.

Coan, J. A., Schaefer, H. S., & Davidson, R. J. (2006). Lending a hand: Social regulation of the neural response to threat. *Psychological Science*, 17, 1032–1039.

Crowe, S. L., & Blair, R. J. R. (2008). The development of antisocial behavior: What can we learn from functional neuroimaging studies? *Developmental Psychopathology, 20*, 1145–1159.

Dennis, T. A., Brotman, L. M., Huang, K. Y., & Gouley, K. K. (2007). Effortful control, social competence, and adjustment problems in children at risk for psychopathology. *Journal of Clinical Child and Adolescent Psychology, 36*, 442–454.

Eisenberger, N. I., Taylor, S. E., Gable, S. L., Hilmert, C. J., & Lieberman, M. D. (2007). Neural pathways link social support to attenuated neuroendocrine stress responses. *Neuroimage, 35*, 1601–1612.

Flinn, M. V., & England, B. G. (1997). Social economics of childhood glucocorticoid stress response and health. *American Journal of Physical Antropology, 102*, 33–53.

Fogel, A. (1993). *Developing through relationships: Communication, self, and culture in early infancy.* Hemel Hempstead, UK: Harvester-Wheatsheaf.

Gailliot, M. T., & Baumeister, R. F. (2007). The physiology of willpower: Linking blood glucose to self-control. *Personality and Social Psychology Review, 11*, 303–327.

Gross, J. J., & Thompson, R. A. (2007). Emotion regulation: Conceptual foundations. In J. J. Gross (Ed.), *Handbook of emotion regulation* (pp. 3–24). New York: Guilford Press.

Haga, S. M., Kraft, P., & Corby, E.-K. (2009). Emotion regulation: Antecedents and well-being outcomes of cognitive reappraisal and expressive suppression in cross-cultural samples. *Journal of Happiness Studies, 10*, 271–291.

Hill, A. L., Degnan, K. A., Calkins, S. D., & Keane, S. P. (2006). Profiles of externalizing behavior problems for boys and girls across preschool: The roles of emotion regulation and inattention. *Developmental Psychology, 42*, 913–928.

Hofer, M. A. (2006). Psychobiological roots of early attachment. *Current Directions in Psychological Science, 15*, 84–88.

House, J. S., Landis, K. R., & Umberson, D. (1988). Social relationships and health. *Science, 241*, 540–545.

Insel, T. R., & Fernald, R. D. (2004). How the brain processes social information: Searching for the social brain. *Annual Review of Neuroscience, 27*, 697–722.

Job, V., Dweck, C. S., & Walton, G. M. (2010). Ego depletion—is it all in your head?: Implicit theories about willpower affect self-regulation. *Psychological Science, 21*, 1686–1693.

Kirsch, P., Esslinger, C., Chen, Q., Mier, D., Lis, S., Siddhanti, S., et al. (2005). Oxytocin modulates neural circuitry for social cognition and fear in humans. *Journal of Neuroscience, 25*, 11489–11493.

Knox, S. S., Adelman, A., Ellison, C. R., Arnett, D. K., Siegmund, K. D., Weidner, G., et al. (2000). Hostility, social support, and carotid artery atherosclerosis in the National Heart, Lung, and Blood Institute Family Heart Study. *American Journal of Cardiology, 86*, 1086–1089.

Krebs, J. R., & Davies, N. B. (1993). *An introduction to behavioural ecology* (3rd ed.). Malden, MA: Blackwell.

Kudo, H., & Dunbar, R. I. M. (2001). Neocortex size and social network size in primates. *Animal Behaviour, 62*, 711–722.

Kurzban, R. (2010). Does the brain consume additional glucose during self-control tasks? *Evolutionary Psychology, 8*, 245–260.

Lewis, M., & Ramsay, D. S. (1999). Effect of maternal soothing on infant stress response. *Child Development, 70*, 11–20.

Lutgendorf, S. K., Sood, A. K., Anderson, B., McGinn, S., Maiseri, H., Dao, M., et al. (2005). Social support, psychological distress, and natural killer cell activity in ovarian cancer. *Journal of Clinical Oncology, 23*, 7105–7113.

Lykins, E. L. B., & Baer, R. A. (2009). Psychological functioning in a sample of long-term practitioners of mindfulness meditation. *Journal of Cognitive Psychotherapy: An International Quarterly, 23*, 226–241.

Moak, Z. B., & Agrawal, A. (2010). The association between perceived interpersonal social support and physical and mental health: Results from the national epidemiological survey on alcohol and related conditions. *Journal of Public Health, 32*, 191–201.

Muraven, M., Collins, R. L., & Nienhaus, K. (2002). Self-control and alcohol restraint: An initial application of the self-control strength model. *Psychology of Addictive Behaviors, 16*, 113–120.

Ochsner, K. N., Bunge, S. A., Gross, J. J., & Gabrieli, J. D. (2002). Rethinking feelings: An FMRI study of the cognitive regulation of emotion. *Journal of Cognitive Neuroscience, 14*, 1215–1229.

Ochsner, K. N., & Gross, J. J. (2005). The cognitive control of emotion. *Trends in Cognitive Science, 9*, 242–249.

Panksepp, J. (1998). *Affective neuroscience: The foundations of human and animal emotions.* New York: Oxford University Press.

Panksepp, J., Nelson, E., & Siviy, S. (1994). Brain opioids and mother–infant social motivation. *Acta Paediatrica, 397*, 40–46.

Proffitt, D. R. (2006). Embodied perception and the economy of action. *Perspectives on Psychological Science, 1*, 110–122.

Rydell, A.-M., Berlin, L., & Bohlin, G. (2003). Emotionality, emotion regulation, and adaptation among 5- to 8-year-old children. *Emotion, 3*, 30–47.

Sbarra, D. A., & Hazan, C. (2008). Co-regulation, dysregulation, self-regulation: An integrative analysis and empirical agenda for understanding adult attachment, separation, loss, and recovery. *Personality and Social Psychology Review, 12*, 141–167.

Smyth, J., & Arigo, D. (2009). Recent evidence supports emotion regulation interventions for improving health in at-risk and clinical populations. *Current Opinion in Psychiatry, 22*, 205–210.

Suveg, C., Sood, E., Comer, J. S., & Kendall, P. C. (2009). Changes in emotion regulation following cognitive-behavioral therapy for anxious youth. *Journal of Clinical Child and Adolescent Psychology, 38*, 390–401.

Taylor, S. E. (2006). Tend and befriend: Biobehavioral bases of affiliation under stress. *Current Directions in Psychological Science, 15*, 273–277.

Tice, D. M., Baumeister, R. F., Shmueli, D., & Muraven, M. (2007). Restoring the self: Positive affect helps improve self-regulation following ego depletion. *Journal of Experimental Social Psychology, 43*, 379–384.

Tomasello, M. (2009). *Why we cooperate*. Cambridge, MA: MIT Press.

Turner-Cobb, J. M., Sephton, S. E., Koopman, C., Blake-Mortimer, J., & Spiegel, D. (2000). Social support and salivary cortisol in women with metastatic breast cancer. *Psychosomatic Medicine*, 62, 337–345.

Uchino, B. N. (2006). Social support and health: A review of physiological processes potentially underlying links to disease outcomes. *Journal of Behavioral Medicine, 29*, 377–387.

Uchino, B. N., Holt-Lunstad, J., Uno, D., Betancourt, R., & Garvey, T. S. (1999). Social support and age-related differences in cardiovascular function: An examination of potential mediators. *Annals of Behavioral Medicine, 21*, 135–142.

CHAPTER 6

Attachment, Commitment, and Relationship Maintenance

When Partners Really Matter

SISI TRAN

University of Toronto Scarborough

JEFFRY A. SIMPSON

University of Minnesota

Imagine two people, Tom and Sarah, who are involved in a romantic relationship. While growing up, Tom had a tumultuous and rocky relationship with his parents, both of whom paid little attention to him and neither of whom Tom felt he could ever fully please. His romantic partner, Sarah, on the other hand, had a good relationship with both of her parents. Sarah felt especially close to her parents and often sought them out for support when she needed it.

When Tom and Sarah started dating, Tom did not want to replicate the difficult relationships he had with his parents, but he worried that he might not live up to Sarah's expectations of him or their relationship. Very early in their relationship, Sarah began to sense Tom's insecurity. She noted that he got anxious whenever she talked with other guys she knew or said anything but positive things about Tom and their relationship. When these situations occurred Tom would immediately get upset and start arguments that often resulted in hurt feelings on both sides. Recognizing this pattern and the likely sources of Tom's insecurity, Sarah began changing how she interacted with Tom. For example, she started to steer Tom away from situations or events that might trigger his concerns and worries. When such events could not be avoided, Sarah did everything she could to accept Tom unconditionally, quickly deescalate conflicts, and reassure him that she loved him and was strongly committed to their relationship. Gradually, Tom's worries abated, and he became less insecure about both himself and the relationship. He worried much less about whether he would "measure up," and when occasional arguments arose he reacted in a more deliberate and constructive manner toward Sarah. Two years later, they were married.

This scenario portrays a set of interpersonal dynamics that occurs fairly often in relationships but has rarely been examined systematically by relationships researchers. In many romantic relationships, at least one partner is likely to have an insecure attachment history. Although insecure histories make people vulnerable to experiencing negative relationship outcomes (Mikulincer & Shaver, 2007), insecure people and their partners are by no means destined to this fate. As the Tom and Sarah scenario illustrates, individuals may find ways to help their insecure partners function better and experience greater satisfaction in relationships. They most likely do so by persistently quelling or disconfirming the chronic worries harbored by their insecurely attached partners, especially when their partners are distressed or feel threatened. Certain individuals, in other words, may buffer or shield their insecure partners from encountering poor relationship outcomes, helping them to think, feel, and behave in more constructive and adaptive ways. One route through which they might do so is by helping insecure partners identify, regulate, and cope more effectively with negative affect, especially in situations that could threaten the relationship itself.

Relationship researchers yearn to understand interpersonal dynamics and outcomes such as these. Most relationship research to date, however, has examined individuals in relationships rather than dyads per se and has treated the individual as the primary unit of analysis. For example, research questions have typically been aimed at examining how Tom's insecure attachment history makes him vulnerable to fears of abandonment or feelings of inadequacy or how Sarah's commitment to the relationship provides the impetus for stronger pro-relationship behaviors. Surprisingly little research has focused on dynamic interaction *between partners*. This chapter focuses on these underexplored interpersonal dynamics. We begin by discussing the important role that constructive emotion regulation has for personal well-being and, in all likelihood, for relationship well-being as well. We then discuss how core principles from two major relationship theories—attachment theory and interdependence theory—can be integrated to explain how and why insecurely attached partners—especially highly anxious ones—should benefit from having romantic partners who are strongly committed to them and the relationship. Following this, we review the results of a recent social interaction study of long-term romantic couples (Tran & Simpson, 2009). This study highlights some of the conditions under which people who are anxiously attached are buffered by partners who are highly committed when couples discuss important accommodative dilemmas—potentially contentious points of disagreement—in their relationship. We conclude the chapter by discussing how these findings advance our understanding of the way "partner buffering" may operate in romantic relationships.

Emotion Regulation and Well-Being

One of the best predictors of personal well-being is the ability to regulate emotions constructively, particularly during difficult or stressful events (Lazarus & Folkman, 1984). Individuals who habitually use constructive, problem-focused modes of coping when troubles arise are generally better at regulating and dampening negative emotions than those who do not. This ability, in turn, is associated with a wide range of positive personal outcomes including better health and greater subjective well-being (see Vohs & Finkel, 2006). Unfortunately, little is known about how constructive emotion regulation translates into adaptive relationship functioning.

Two major lines of research, however, have indirectly addressed this issue. Research on the personality trait of neuroticism has shown that people who report being more emotionally unstable (i.e., highly neurotic) have much less satisfying relationships that are more likely to end in dissolution or divorce (Karney & Bradbury, 1995). Similarly, individuals with insecure attachment styles (i.e., who enact either emotion-focused coping strategies in the case of anxious attachment styles or avoidance coping strategies in the case of avoidant attachment styles) also experience poorer relationship outcomes (Mikulincer & Florian, 1998). Emotion-focused coping entails the use of tactics such as vigilantly focusing on, ruminating about, and amplifying the source, severity, or chronicity of distress. Avoidance coping entails the use of tactics such as denying, ignoring, discounting, and failing to acknowledge the existence of stress along with the negative effects it has on the self. People who use one or both types of insecure coping strategies generally experience less interdependence, less trust, more emotional negativity, and less satisfying relationships than people who use more secure, problem-focused coping strategies (Simpson, 1990). Problem-focused coping consists of tactics that directly address and eventually "solve" the problem or issue that is causing distress, which allows securely attached people to resume other important activities without having to use vigilance tactics (in the case of highly anxious persons) or defensive tactics (in the case of highly avoidant persons).

At this point, very little is known about whether or how romantic partners "regulate" each other when stressful relationship events might be and are encountered. We propose that the ability to skillfully regulate the emotions of not only oneself but also one's *partner* may be one of the most important assets that partners can bring to a relationship. In this way, individuals can serve as an important "resource" when their romantic partners experience negative emotions by providing needed emotional or instrumental support.

Attachment Theory

Within the relationships literature, attachment theory (Bowlby, 1969/1982, 1973, 1980) has become one of the most generative and influential metatheories. According to this theory, patterns of interaction with attachment figures (such as primary caregivers) that occur early in life shape an individual's beliefs and expectations of later relationships (Bowlby 1969/1982, 1973, 1980). Once formed, these relationship expectations or "working models" gradually lead individuals to develop specific attachment orientations. Two orthogonal dimensions underlie adult attachment orientations (Brennan, Clark, & Shaver, 1998; Simpson, Rholes, & Phillips, 1996). The first dimension, labeled *anxiety*, taps concerns that relationship partners might not be available and supportive when needed and that love may not be fully reciprocated. The second dimension, termed *avoidance*, indexes the desire to limit intimacy and dependence and to maintain comfortable psychological and emotional independence from relationship partners. Individuals who score low on both attachment dimensions are prototypically "secure" in that they feel comfortable with closeness and intimacy and are confident in the availability and benevolent intentions of their partners.

Anxious attachment develops from receiving inconsistent or unpredictable care from prior attachment figures (Cassidy & Berlin, 1994). The anxious orientation is defined by concerns about one's worthiness of love, which is manifested in chronic fear of rejection and doubts about the ultimate availability and supportiveness of attachment figures. Highly anxious individuals are hypervigilant with respect to the availability of support from their partners, and they ruminate over worst-case relationship outcomes (Kobak & Sceery, 1988; Mikulincer, Florian, & Weller, 1993). As a consequence, they use hyperactivation strategies, which include clinging, controlling, and coercive behaviors, to ensure that their attachment figures remain psychologically close and available (Cassidy & Kobak, 1988; Mikulincer & Shaver, 2007).

Avoidant attachment, in contrast, emerges from a history of unsuccessful bids for proximity in which an individual's efforts are consistently met with neglect or rejection from attachment figures (Bowlby, 1973; Crittenden

& Ainsworth, 1989). For highly avoidant individuals, the attachment system is triggered by reminders of their futile efforts to solicit care and support, making them vulnerable to reexperiencing emotional rejection (Bartholomew, 1990). As a result, highly avoidant people use defensive deactivation strategies that limit intimacy and deny or suppress their latent needs for greater closeness to attachment figures (Bowlby, 1980; Cassidy & Kobak, 1988; Crittenden & Ainsworth). Avoidant attachment is also characterized by strong preferences to create and maintain autonomy, control, and emotional distance in interpersonal contexts (Fraley, Davis, & Shaver, 1998; Fraley & Shaver, 1998; Mikulincer, 1998; Shaver & Hazan, 1993).

Interdependence Theory

Interdependence theory (Kelley & Thibaut, 1978; Thibaut & Kelley, 1959) represents a second major theoretical framework within the study of relationships. This theory suggests that most individuals undergo a "transformation of motivation" when deciding whether to do something that is good for themselves versus something that is good for their partners or relationships (Kelley & Thibaut, 1978). According to this theory, a distinction must be made between the given matrix and the effective matrix (Figure 6.1). The given matrix represents an individual's primitive or "gut-level" self-centered preferences when a problem is encountered. People generally experience negative emotions when treated badly,

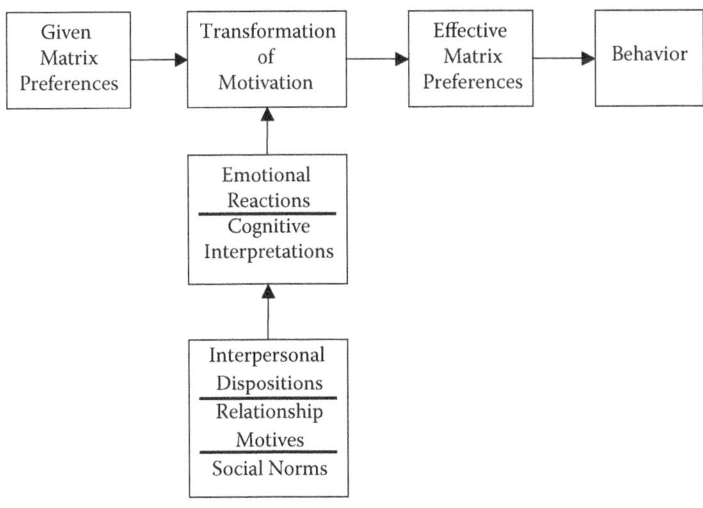

Figure 6.1 The transformation of motivation model (From Rusbult, C.E., Yovetich, N. A., & Verette, J., In *Knowledge Structures in Close Relationships: A Social Psychological Approach*, Hillsdale, NJ, Lawrence Erlbaum Associates, Inc., 1996.).

and their immediate impulse often is to reciprocate negative behavior in kind. Reactions indexed by the given matrix, however, do not necessarily dictate how an individual actually behaves when confronted with partner negativity. According to interdependence theory, most individuals undergo a transformation of motivation when deciding whether to act on their initial, self-interested preferences (e.g., to retaliate) or whether to behave in ways that might promote broader relationship goals (e.g., to find constructive ways to resolve the problem; see Rusbult, Arriaga, & Agnew, 2002). Determinants of transformation tendencies include interpersonal orientations, such as individuals' dispositional tendencies or relationship goals and motives. These variables are believed to determine the amount of transformation that occurs via their impact on cognitive interpretations of and emotional reactions to the specific situation in which relationship partners' self-interests are at odds with their broader relationship goals. In other words, the transformation process model proposes that the regulation of one's thoughts and emotions results in more adaptive behaviors during accommodative dilemmas. The effective matrix, therefore, reflects the eventual transformation of the given matrix (if transformation occurs), and it ultimately guides how individuals behave toward their partners.

According to Rusbult and colleagues (Rusbult, Verette, Whitney, Slovik, & Lipkus, 1991), an individual's willingness to respond constructively and inhibit impulses to react destructively when a partner displays potentially destructive behaviors defines accommodation. Commitment is believed to be the most important construct for understanding motivations that eventually produce accommodation (Rusbult et al., 1991). Commitment represents concern for the future and the stability of the relationship plus the desire for the relationship to continue. Commitment, therefore, correlates highly with persistence in relationships (Bui, Peplau, & Hill; 1996; Drigotas & Rusbult, 1992; Drigotas, Rusbult, & Verette, 1999; Etcheverry & Le, 2005), and it is the strongest predictor of most pro-relationship maintenance behaviors, such as derogation of alternatives (Johnson & Rusbult, 1989) and willingness to make sacrifices for the partner or relationship (Van Lange et al., 1997). Greater commitment also predicts the enactment of more constructive behaviors and fewer destructive ones when partners' interests are not perfectly aligned (Campbell & Foster, 2002; Etcheverry & Le; Menzies-Toman & Lydon, 2005; Rusbult, Bissonnette, Arriaga, & Cox, 1998; Rusbult et al., 1991; Rusbult, Yovetich, & Verette, 1996; Wieselquist, Rusbult, Foster, & Agnew, 1999).

A Dyadic View of Attachment and Commitment

For some individuals, a history of negative interpersonal experiences may prevent them from behaving in ways that could bolster the stability and longevity of their relationships. In response to previous maladaptive relationships, people are likely to develop negative or unrealistic expectations about the availability, responsiveness, and intentions of romantic partners (Baldwin, 1992). Patterns of negativity may thus be maintained in current relationships via behavioral confirmation processes (Snyder & Stukas, 1999). For example, insecurely attached individuals may anticipate negative reactions or behaviors from their romantic partners, perceive greater partner negativity or malintent, overreact to these perceptions, and, as a result, unwittingly elicit negative behaviors from their partners. Indeed, women who are more rejection sensitive (and who also tend to be more anxiously attached) behave in a more hostile and defensive manner during conflict interactions, which leads their romantic partners to experience greater anger and dissatisfaction (Downey, Freitas, Michaelis, & Khouri, 1998). Negative expectations and relationship insecurities, therefore, can easily subvert relationship quality.

Fortunately, not all relationships in which one or both partners are insecurely attached (or hold negative relationship expectations) are destined for failure. Highly anxious people who perceive higher levels of support from their spouses report better marital functioning (Rholes, Simpson, Campbell, & Grich, 2001) and believe they will have better future relationship outcomes (Campbell, Simpson, Boldry, & Kashy, 2005). Insecurely attached people involved in highly committed relationships might be able to quell or suspend their worries about rejection and loss, eventually extricating themselves from a continuing cycle of negative thoughts, feelings, and behaviors. Campbell, Simpson, Kashy, and Rholes (2001) also found that the negative effects generally observed for people with greater attachment insecurities are attenuated when insecure individuals are more dependent on their partners. In other words, greater dependence or commitment may provide insecurely attached people with a broader, long-term perspective that might help them achieve happier and more stable relationships (Kelley, 1983). This motivation to sustain the relationship might allow them to disregard or sidestep their immediate attachment-based concerns and worries and work more effectively toward meeting their long-term relationship goals. Greater relationship commitment, in other words, may effectively buffer attachment insecurities.

However, the *partner's* level of commitment should have an even stronger effect on an individual's emotions and behavioral reactions to relationship-threatening events, given that partners can easily destabilize and terminate relationships (Attridge, Berscheid, & Simpson, 1995). Indeed, greater commitment by *partners* may be the foundation upon which insecurely attached individuals can feel more confident that their partners truly love, care for, and respect them. Rather than being wrapped up in feelings of vulnerability and insecurity, this realization may allow insecure people to experience less intense negative affect and better regulate their emotions, which could in turn enable them to behave in a more constructive, accommodating manner when relationship-threatening events are encountered.

Less committed individuals, in contrast, should experience more negative outcomes, especially if they are involved with highly insecure partners. The combination of low personal commitment and high partner insecurity should culminate in particularly negative outcomes in terms of how less committed people think, feel, and behave in relationship-threatening situations. The maladaptive coping strategies characteristic of highly insecure individuals, in other words, may be even worse for the relationship if one or both partners lack the commitment and positive motivation necessary to counteract these tendencies.

The buffering effects of commitment, however, should be stronger for more anxiously attached than for more avoidantly attached people. Avoidantly attached individuals are motivated to create and maintain control and sufficient emotional distance in their relationships (Mikulincer, 1998). Without sufficient control and autonomy, highly avoidant people may feel vulnerable and even "trapped" in relationships. As a consequence, higher levels of one's own commitment or having partners who are highly committed may threaten highly avoidant individuals' need for autonomy and control. In contrast, highly anxious individuals are motivated to achieve greater security and reassurance from their partners (Mikulincer). For this reason, greater self-commitment *and especially greater partner commitment* may reduce relationship threat and allow highly anxious people to believe they are closer to achieving sufficient feelings of security.

According to transformation of motivation principles (Kelley & Thibaut, 1978), individuals' interpersonal dispositions (e.g., attachment anxiety) and relationship motives (e.g., the desire to maintain the current relationship) should affect their perceptual and emotional responses to important relationship events (e.g., an accommodative dilemma in the relationship). These thoughts and emotions, in turn, should affect whether individuals behave in an accommodating manner, especially during a relationship-threatening interaction. The specific behaviors individuals enact, however, should be more strongly influenced by the specific thoughts and feelings that they have during a threatening interaction than by their global dispositions or motives. For example, the deep-seated insecurities of highly anxious individuals may be manifested in intense negative emotional responses to a threatening situation, which results in hostile or defensive behaviors. However, the countervailing desire to maintain the current relationship may help these individuals sidestep their immediate, gut impulse to behave defensively by transforming their perceptions or interpretations of the threatening situation. This, in turn, may permit highly anxious individuals to regulate their emotions more constructively and respond to their partners in a more adaptive and benevolent manner.

A Study of Attachment, Commitment, and Accommodation

To test these predictions, we conducted a videotaped social interaction study (see Tran & Simpson, 2009). A total of 74 married couples in the Minneapolis-St. Paul area participated in the study. The mean length of marriage was 5 years, and the mean ages of men and women were 32 and 33, respectively. Of the participants, 126 were Caucasian, 3 were African American, 7 were Hispanic, 6 were American Indian, and 6 were Asian.

Participants first completed a set of questionnaires privately and independently of their spouse. An adapted version of the Experiences in Close Relationships Scale (ECR; Brennan et al., 1998) was used to assess the two adult attachment dimensions (anxiety and avoidance). The anxiety dimension assesses the degree to which individuals have negative views of themselves as relationship partners and are preoccupied with abandonment and loss of attachment figures. Sample items from the anxiety scale are, "I worry about being abandoned," and, "I find that romantic partners don't want to get as close as I would like." The avoidance dimension taps the degree to which individuals harbor negative views of others and seek to avoid closeness and intimacy in relationships. Sample items from the avoidance scale are, "I find it difficult to allow myself to depend on romantic partners," and "I am very comfortable being close to romantic partners" (reverse scored). Relationship commitment was assessed using the Investment Model Commitment Scale (Rusbult, 1983). Sample items include, "How much longer do you want your current relationship to last?" and "Do you feel committed to maintaining your relationship with your partner?"

Each couple then engaged in two videotaped accommodative dilemma discussions. In the first dilemma, one partner (the initiator) was randomly assigned to initiate a discussion about a characteristic, habit, or behavior of the other partner that the initiator wanted to see change. Accommodative dilemmas are a particularly good context in which to test transformation of motivation processes because the partner (the accommodator) has the option to react constructively (by attempting to accommodate the request for change), to react neutrally, or to react destructively (in line with personal self-interests). The initiator and accommodator roles were reversed in the second accommodative dilemma discussion.

Immediately following each discussion, a self-report measure assessed attachment-related feelings of acceptance (e.g., loved, supported, cared for, comforted, secure) and rejection (e.g., dismissed, abandoned, hostile, rejected, insecure) during the discussion. A composite variable for emotional reactions was computed with positive scores signifying more positive emotions and negative scores signifying more negative emotions.

To assess participants' constructive and destructive behaviors, each videotaped interaction was independently rated by five trained researchers. The coding scheme was developed based on Rusbult and Zembrodt's (1983) dimensions of constructive and destructive behaviors. Specifically, each coder rated the target partners (the accommodators) in terms of the extent to which they displayed constructive behaviors (e.g., compromising, suggesting solutions, showing optimism, attempting to resolve the problem) and destructive behaviors (e.g., criticizing their partner, using a condescending tone, allowing the problem to continue, avoiding the issue). The composite variable for accommodative behaviors was composed of ratings of constructive and destructive behaviors. Positive scores reflected the enactment of more constructive behaviors, and negative scores reflected the enactment of more destructive behaviors.

Primary Results

Descriptive Analyses at the Individual Level

In each interaction, only one person was assigned to initiate the accommodative dilemma topic for discussion, and the other person was allowed to respond. Although both partners participated in each discussion, the measures of emotional and behavioral reactions for each individual were assessed from the discussion in which the individual was the responder (accommodator). A summary of the zero-order correlations between all major variables are shown in Table 6.1. As expected, women's and men's anxiety and avoidance scores were negatively correlated with adaptive emotional responses during the discussions. In other words, more insecurely attached individuals felt greater rejection and less acceptance during their interactions. Women's anxiety and avoidance scores were also negatively associated with observer-rated accommodative behaviors. Specifically, highly anxious and highly avoidant women displayed fewer constructive behaviors, and highly avoidant women displayed more destructive behaviors.

In contrast to the effects for attachment insecurity, women who were more committed to their partners and relationships reported feeling greater acceptance and less rejection during their discussions. Additionally, women's commitment was positively associated with their accommodative behaviors. Because there were nonsignificant correlations between men's commitment and their emotional and behavioral outcomes, differences between the correlations for men's commitment and women's commitment were tested. Compared with men's commitment, women's commitment

Table 6.1 Correlations Among All Major Variables

	1	2	3	4	5	6	7	8	9
Women's anxiety									
Women's avoidance	.38**								
Women's commitment	–.30**	–.50**							
Women's emotional reactions	–.33**	–.31**	.44**						
Women's behavioral reactions	–.24**	–.26**	.41**	.52**					
Men's anxiety	.17	.28*	–.21†	–.06	–.29*				
Men's avoidance	.10	.16	–.14	–.11	.02	.28*			
Men's commitment	–.03	–.03	.22†	.06	.11	–.07	–.44*		
Men's emotional reactions	–.27*	–.31**	.39**	.44**	.48**	–.26*	–.25*	.16	
Men's behavioral reactions	–.20†	–.21†	.45**	.54**	.43**	–.18	–.12	.16	.43**

Note: $N = 74$ women, 74 men.
† $p < .10$.
* $p < .05$.
** $p < .01$.

was significantly more strongly related to their own emotional experiences during the discussions and with the constructive and destructive behaviors they enacted. Given these gender differences, we model and discuss them in greater detail in the primary analyses.

Correlations between partners (i.e., within-couple correlations) showed that women's anxiety and avoidance scores were negatively associated with men's emotional responses during the discussion, and women's relationship commitment scores were positively correlated with men's emotional and behavioral responses. Interestingly, men's anxiety, avoidance, and commitment scores revealed considerably fewer significant associations with women's outcomes. Not surprisingly, however, women's and men's emotional and behavioral responses to the discussion were closely linked.

Primary Analyses at the Dyadic Level

The primary analyses were conducted using the Actor-Partner Interdependence Model (APIM; Kenny, Kashy, & Cook, 2006). The APIM is appropriate for use when the dyad (the romantic couple) is the unit of analysis and tests are performed between and within dyads (Kashy & Kenny, 2000). The APIM tests not only whether actors' own attributes predict their responses and behaviors, controlling for their partner's attributes, but also whether their partner's attributes predict their responses and behaviors, controlling for their attributes. For example, in the current study, an actor effect for attachment anxiety would be evident if individuals' scores on the anxiety dimension predicted their destructive behaviors, controlling for their partner's level of anxiety. A partner effect would be evident if actors' *partner's* anxiety score predicted their destructive behaviors, controlling for their own level of anxiety.

Similar to previous analyses, persons were labeled differently depending on their role in each of the discussions. Specifically, measures from the target person (the individual in the role of responder or accommodator) were coded as "actor" variables, and measures from the other person (the individual in the role of discussion initiator) were coded as "partner" variables. Thus, each couple member was an "actor" in only one discussion. For our analyses, actor anxiety, actor avoidance, partner anxiety, and partner avoidance scores were entered as the first block of predictor variables; actor and partner commitment scores and actor gender were entered in the second block; the two-way interactions between actor anxiety × actor commitment, actor avoidance × actor commitment, partner anxiety × partner commitment, and partner avoidance × partner commitment were entered in the third block; and actor anxiety × partner commitment, actor avoidance × partner commitment, partner anxiety × actor commitment, and partner avoidance × actor commitment were entered in the final block.

Attachment by Commitment Effects

As hypothesized, we found significant interactions between partner anxiety and actor commitment predicting both emotional reactions and accommodative behaviors during the discussions. These interactions, which are depicted in Figures 6.2a and 6.2b, indicate that people married to less anxious partners revealed no significant difference in their emotional reactions and behavioral accommodation during the discussions, regardless of their degree of commitment to the relationship. These findings suggest that being married to less anxious (or more secure) partners allowed individuals to maintain a more positive set of emotional and behavioral reactions to the discussions, regardless of their own level of commitment.

(a)

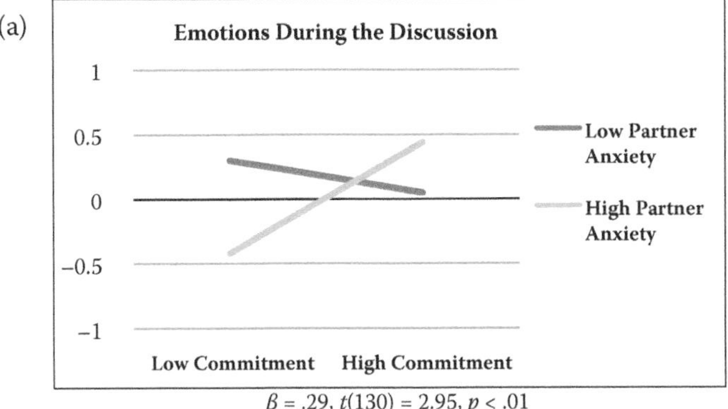

$\beta = .29, t(130) = 2.95, p < .01$

(b)

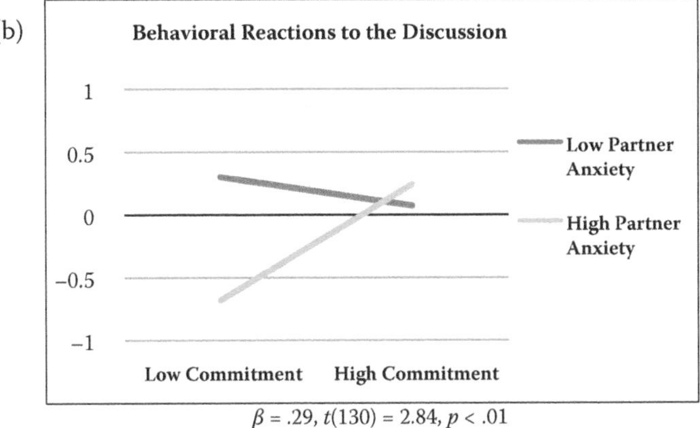

$\beta = .29, t(130) = 2.84, p < .01$

Figure 6.2 (a) The 2-way interaction between actor commitment and partner attachment anxiety predicting emotional reactions to the discussion. (b) The 2-way interaction between actor commitment and partner attachment anxiety predicting behavioral reactions to the discussion. All of the variables are centered. Regression lines are plotted for individuals scoring 1 standard deviation above and below the sample means on anxiety and commitment.

However, individuals who were involved with highly anxious partners experienced more negative emotions and behaved more destructively, but only if they were less committed to the relationship. Being married to a highly anxious spouse likely requires the display of persistent reassurance to calm and abate insecure spouses' worries and insecurities. It may be difficult to engage in and sustain effective reassurance, particularly if one is less committed to the relationship. However, if individuals who are involved with highly anxious partners reported greater relationship commitment, they experience more positive emotions and display more accommodative behaviors during the accommodative discussions. Greater

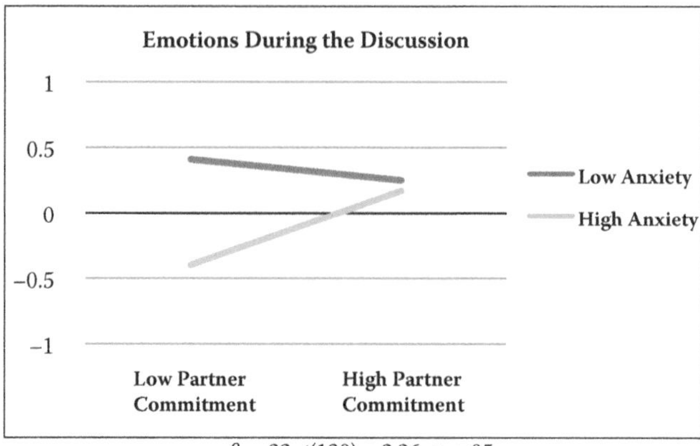

$\beta = .22$, $t(130) = 2.26$, $p < .05$

Figure 6.3 The 2-way interaction between actor attachment anxiety and partner commitment predicting emotional reactions to the discussion. All of the variables are centered. Regression lines are plotted for individuals scoring 1 standard deviation above and below the sample means on anxiety and commitment.

commitment, in other words, appears to curtail or cushion the potentially negative emotions and destructive behaviors that can arise when interacting with highly anxious partners in relationship-threatening situations.

Moreover, the inverse relation between *individuals'* attachment anxiety and their emotional reactions during the discussions was moderated by their *partners'* commitment to the relationship. As depicted in Figure 6.3, highly anxious individuals reported more negative emotions than their less anxious counterparts did, particularly when their partners reported being less committed to the relationship. In other words, being married to a less committed partner appears to exacerbate feelings of insecurity in highly anxious individuals. If, however, their partners report being more committed, highly anxious individuals report comparatively fewer negative emotions. Greater partner commitment, therefore, seems to diminish highly anxious people's negative reactions during potentially relationship-threatening interactions.

A Process Model for Dyadic Gender Effects

Relative to men, women's commitment was significantly more strongly associated with their emotional responses and with the constructive and destructive behaviors they displayed during the discussions. Given these gender differences, we developed and tested a process model of relations between wives' commitment, husbands' commitment, their respective reports of emotional reactions during their discussion, and their respec-

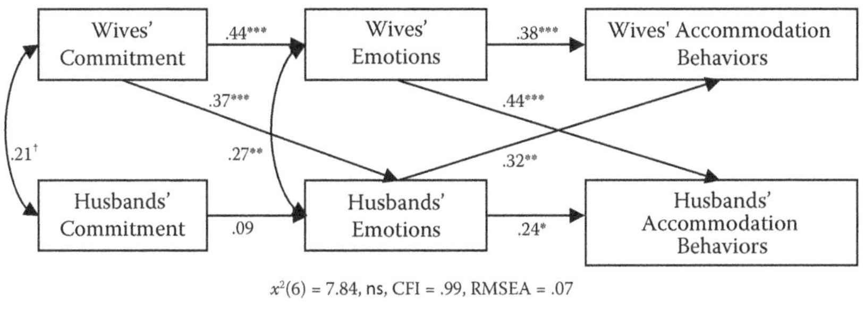

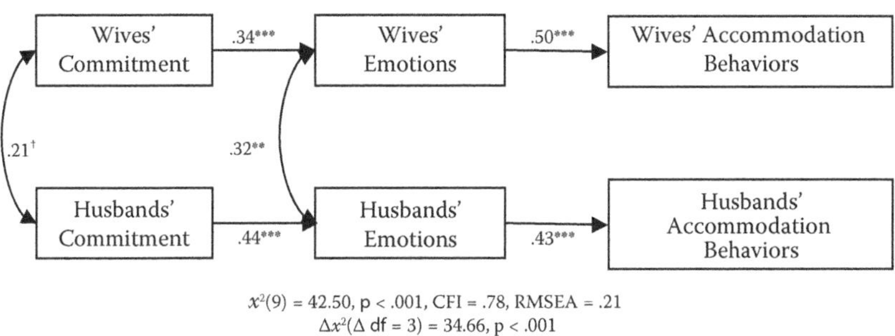

Figure 6.4 (a) The dyadic process model for the mediating role of emotions. (b) The comparison model without the dyadic paths.

tive behavioral reactions during the discussion. As shown in Figure 6.4a, this model fit the data reasonably well.

As expected, the link between wives' commitment and their own accommodative behaviors was mediated by their emotional reactions. This outcome supports a core theoretical proposition made by Kelley and Thibaut (1978). Specifically, greater commitment to the relationship was associated with having more positive emotions during a potentially relationship-threatening event, and these positive emotions in turn predicted the enactment of more constructive behaviors. Consistent with the transformation of motivation process model (see Figure 6.1), more accommodative behaviors were due at least in part to relationship-enhancing motives (assessed by wives' commitment to the partner or relationship) as mediated through the effective control or suppression of potentially harmful emotional reactions during the accommodative dilemma discussions.

Although husbands' emotional reactions were significantly associated with their accommodative behaviors, husbands' level of commitment was not significantly associated with their emotional reactions during the discussion. Indeed, wives' degree of commitment had a stronger impact on husbands' emotional reactions than did husbands' own reported levels of commitment. These findings imply that there could be a slight disconnect between men's level of commitment and their expression of commitment. Thus, despite the fact that a man is highly committed to his current relationship, he may not necessarily communicate that devotion to his spouse via his emotions and behaviors. Women, by comparison, may express their thoughts and feelings more openly and more directly. These findings suggest that wives' level of commitment may play a stronger role in determining how their husbands feel and behave during threatening interactions, independent of how committed their husbands report being.

Interestingly, wives' degree of commitment predicted their husbands' emotional reactions during the discussions, and husbands' emotional reactions predicted their wives' discussion behaviors. The association between wives' commitment and their accommodative behaviors, in other words, was partially mediated by their *husbands'* emotional responses during the discussions, highlighting the dynamic interchange between partners. Consistent with these dyadic effects, wives' emotional reactions also predicted their husbands' discussion behaviors. These findings reveal that individuals' behaviors are impacted not only by their own motivational and emotional underpinnings but also by their *partners'* reactions to accommodative dilemmas.

The large correlations between husbands' and wives' emotions and behaviors (*r*'s range from .54 to .79) indicate that partners' reactions were closely linked. Not only did husbands and wives influence one another's responses; each spouse's own emotional and behavioral reactions had a reciprocal influence as well. Although the transformation of motivation model (Kelley & Thibaut, 1978) offers one explanation for the mediating role of emotion regulation in predicting more accommodative behaviors, a second alternative model was tested to examine the mediating role of constructive behaviors on the relation between partners' relationship commitment and their emotional reactions. Similar to the first model, this alternative model (shown in Figure 6.5a) also revealed an adequate fit to the data, suggesting a mutually influential role between emotional and behavioral responses. Although the present cross-sectional study cannot test for a causal relation between these variables, it is clear that there is a strong dyadic link between the way partners feel and the way they behave in their relationships. Framed another way, the control and suppression of potentially harmful emotional responses from one or both partners most likely led to more accommodating behaviors; conversely, more constructive behaviors displayed by one or both partners most likely produced more positive emotional responses.

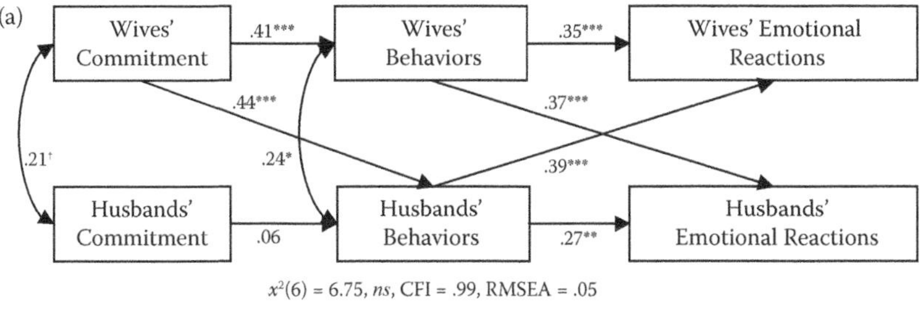

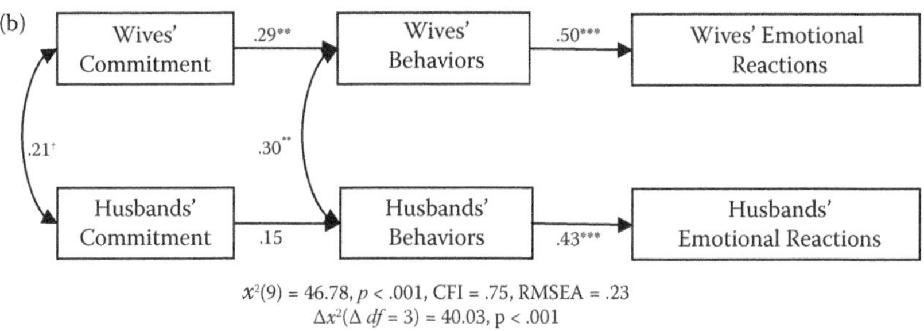

Figure 6.5 (a) The dyadic process model for the mediating role of behaviors. (b) The comparison model without the dyadic paths.

Finally, we tested a set of comparison models that excluded the dyadic paths between wives and husbands. As shown in Figures 6.4b and 6.5b, the comparison models fit the data less adequately than did the models that included the dyadic paths. These results attest to the importance of modeling *both* partners when examining relationship dynamics. After all, relationship outcomes depend on the interaction *between* the partners rather than the thoughts and feelings harbored by merely one partner.

Implications

The findings of this study are novel by revealing how certain individual characteristics (e.g., attachment orientations) intersect with important features of a specific relationship (e.g., commitment) to jointly affect how romantic partners feel and behave during an accommodative dilemma discussion. By examining actual relationship dynamics as they unfold between marital partners during a potentially relationship-threatening interaction, we were able to model how individuals' feelings and behaviors were tied to those of their spouse. As expected, we found that more insecurely attached individuals felt greater rejection and less acceptance from their partners during these discussions. Consistent with previous research (e.g., Gaines et al., 1997; Simpson et al., 1996), such persons also displayed fewer constructive and more destructive behaviors. However, individuals' feelings of insecurity appeared to be diminished by their partners' relatively greater commitment to the relationship. Highly anxious individuals' chronic fears of abandonment and their hypervigilance to signs of imminent rejection make them feel particularly vulnerable in relationship-threatening situations. However, when their partners consistently show signs of commitment and motivation to sustain the relationship, these fears seem to be quelled and quieted, resulting in reduced feelings of insecurity and enhanced positive emotions.

Not surprisingly, lower levels of partner commitment merely exacerbated the insecurities harbored by highly anxious individuals. Less committed individuals who were married to highly anxious partners, for instance, felt more negative emotions and behaved more destructively during the accommodative dilemma discussions. In other words, the specific combination of lower self-commitment and higher partner anxiety culminated in the most negative outcomes. Fortunately, greater self-commitment buffered many of the deleterious effects normally associated with greater attachment anxiety in partners. More committed individuals, for example, experienced more positive emotions during the discussions and in turn behaved more constructively, despite having highly anxious partners. Thus, if individuals are involved with highly anxious partners, it is important that they create and sustain higher levels of commitment and then directly express their heightened commitment to counteract the negative effects of their partners' attachment insecurity.

The results also confirmed that both wives' and husbands' emotions and behaviors during the discussion exerted a significant impact on the other's emotional and behavioral responses during the discussion. This reciprocal influence indicates why studying only one partner in a relationship provides insufficient data and information. The current findings make even more sense when one recognizes that the *partner's* reactions to major relationship events often may be the best barometer of how well the relationship is doing (Attridge et al., 1995). Indeed, the effect of the *partners'* responses on each individual's own responses testifies to the importance of dyadic influences in regulating emotional and behavioral experiences in partners.

Although longitudinal effects could not be examined in the current study, the constructs that we examined appeared to have a reciprocal effect on another. As this research highlights, higher levels of relationship commitment were conveyed through more positive emotional reactions during potentially relationship-threatening interactions. These emotions, in turn, predicted the enactment of more constructive as well as less destructive behaviors in both the self and the partner. This pattern of findings attests to the importance of emotion regulation in promoting and perhaps enabling constructive accommodation behaviors. Inversely, the present research also shows that constructive behaviors mediated the link between relationship commitment and emotional responses. Thus, being motivated to sustain the relationship led to greater accommodative behaviors, which then led to more positive emotional responses in both the self and partner. Over time one may begin to see the reciprocal influence of emotional and behavioral reactions even more clearly as the motivation to sustain or improve the relationship establishes a new interaction trajectory that gradually generates enhanced positivity.

Greater commitment may sustain "vulnerable" relationships sufficiently long enough for partners to develop a sense of greater trust in one another, allowing attachment insecurities to gradually wane. Declines in attachment insecurity may then shift how highly anxious individuals react when compromises must be forged with their partners, ultimately resulting in more positive emotional and behavioral reactions. Highly committed individuals, in other words, may diminish their partners' degree of insecurity over time by consistently providing a "secure base," especially in situations where the partners' outcomes are not correspondent (Simpson, 2007). Conversely, highly anxious individuals who begin to feel more secure in their relationship may learn to accept their partners' support and affection fully, coming to believe that they are worthy of love. These feelings of enhanced security may then launch greater commitment.

The current research highlights how individuals' mental representations (working models) presumably forged in earlier relationships operate in conjunction with proximal qualities of interdependence between partners that exists in current relationships. As this research shows, being highly anxious or being involved with a highly anxious partner may initially impede one's inclination to react constructively to relationship-threatening events. Greater relationship commitment from oneself or one's partner, however, functions as a buffer against the potential negative effects of attachment insecurity, diminishing feelings of rejection, enhancing feelings of acceptance, and promoting more constructive accommodative behaviors. The motivation to preserve and stabilize relationships, therefore, can at times override the maladaptive working models and coping strategies harbored by insecure people.

Conclusion

This research showcases the need and value of adopting a dyadic perspective to the study of relationships. Previous investigations of attachment and interdependence phenomena have all too often studied individuals in relationships rather than *partners* in relationships. Studies that focus solely on individuals cannot measure and model the ways partners jointly impact one another. As this research documents, greater commitment on the part of at least one partner buffers the other partner's attachment insecurity. Moreover, each partner's emotional and behavioral reactions also have clear effects on the other's outcomes. The characteristics of *both* partners, therefore, are essential to examining, modeling, and fully understanding relationship phenomena.

The research reviewed in this chapter underscores the need to understand characteristics that exist *within* individuals as well as emergent properties that exist *between* partners if one wants to fully comprehend relationship dynamics. To return to our hypothetical couple, one might anticipate that Tom is likely to have less satisfying and less stable romantic relationships in light of his prior history of attachment insecurity. Sarah's commitment and devotion to Tom, however, ought to quell the potentially negative effects of Tom's working models and attenuate the link between his negative working models and his negative emotional and behavioral reactions when the two disagree or do not see eye to eye. Negative relationship histories can and often do hinder an individual's ability to cope effectively with relationship-threatening events. However, commitment—particularly the *partner's* level of commitment—can offset negative outcomes by curtailing the tendency of insecurely attached people to react negatively and by promoting more constructive responses. Despite their negative relationship histories, individuals who are strongly motivated to overcome their vulnerabilities may be able to realign relationship dynamics to enhance adaptive functioning and eventually attain positive outcomes.

References

Attridge, M., Berscheid, E., & Simpson, J. (1995). Predicting relationship stability from both partners versus one. *Journal of Personality and Social Psychology, 69*, 254–268.

Baldwin, M. W. (1992). Relational schemas and the processing of social information. *Psychological Bulletin, 112*, 461–484.

Bartholomew, K. (1990). Avoidance of intimacy: An attachment perspective. *Journal of Social and Personal Relationships, 7*, 147–178.

Bowlby, J. (1969/1982). *Attachment and loss: Attachment*. New York: Basic Books.

Bowlby, J. (1973). *Attachment and loss: Separation, anxiety, and anger*. New York: Basic Books.

Bowlby, J. (1980). *Attachment and loss: Sadness and depression*. New York: Basic Books.

Brennan, K. A., Clark, C. L., & Shaver, P. R. (1998). Self-report measurement of adult attachment. In J. A. Simpson & W. S. Rholes (Eds.), *Attachment theory and close relationships* (pp. 46–76). New York: Guilford Press.

Bui, K. T., Peplau, L. A., & Hill, C. T. (1996). Testing the Rusbult model of relationship commitment and stability in a 15-year study of heterosexual couples. *Personality and Social Psychology Bulletin, 22,* 1244–1257.

Campbell, L., Simpson, J. A., Boldry, J., & Kashy, D. A. (2005). Perceptions of conflict and support in romantic relationships: The role of attachment anxiety. *Journal of Personality and Social Psychology, 88,* 510–531.

Campbell, L., Simpson, J. A., Kashy, D. A., & Rholes, W. S. (2001). Attachment orientations, dependence, and behavior in a stressful situation: An application of the actor-partner interderpendence model. *Journal of Social and Personal Relationships, 18,* 821–843.

Campbell, W. K., & Foster, C. (2002). Narcissism and commitment in romantic relationships: An investment model analysis. *Personality and Social Psychology Bulletin, 28,* 484–495.

Cassidy, J., & Berlin, L. J. (1994). The insecure/ambivalent pattern of attachment: Theory and research. *Child Development, 65,* 971–991.

Cassidy, J., & Kobak, R. R. (1988). Avoidance and its relationship with other defensive processes. In J. Belsky & T. Nezworski (Eds.), *Clinical implications of attachment* (pp. 300–323). Hillsdale, NJ: Erlbaum.

Crittenden, P. M., & Ainsworth, M. (1989). Child maltreatment and attachment theory. In D. Cicchetti & V. Carlson (Eds.), *Child maltreatment: Theory and research on the causes and consequences of child abuse and neglect* (pp. 432–463). Cambridge, England: Cambridge University Press.

Downey, G., Freitas, A. L., Michaelis, B., & Khouri, H. (1998). The self-fulfilling prophecy in close relationships: Rejection sensitivity and rejection by romantic partners. *Journal of Personality and Social Psychology, 75,* 545–560.

Drigotas, S. M., & Rusbult, C. E. (1992). Should I stay or should I go? A dependence model of breakups. *Journal of Personality and Social Psychology, 62,w* 62–87.

Drigotas, S. M., Rusbult, C. E., & Verette, J. (1999). Level of commitment, mutuality of commitment, and couple well-being. *Personal Relationships, 6,* 389–409.

Etcheverry, P. E., & Le, B. (2005). Thinking about commitment: Accessibility of commitment and prediction of relationship persistence, accommodation, and willingness to sacrifice. *Personal Relationships, 12,* 103–123.

Fan, X., Thompson, B., & Wang, L. (1999). Effects of sample size, estimation methods, and model specification on structural equation modeling fit indexes. *Structural Equation Modeling, 6,* 56–83.

Fraley, R. C., Davis, K. E., & Shaver, P. R. (1998). Dismissing avoidance and the defensive organization of emotion, cognition, and behavior. In J. A. Simpson & W. S. Rholes (Eds.), *Attachment theory and close relationships* (pp. 249–279). New York: Guilford.

Fraley, R. C., & Shaver, P. R. (1998). Airport separations: A naturalistic study of adult attachment dynamics in separating couples. *Journal of Personality and Social Psychology, 75,* 1198–1212.

Gaines, S. O., Reis, H. T., Summers, S., Rusbult, C. E., Cox, C. L., Wexler, M. O., et al. (1997). Impact of attachment style on reactions to accommodative dilemmas in close relationships. *Personal Relationships, 4,* 93–113.

Gallo, L. C., & Smith. T. W. (2001). Attachment style in marriage: Adjustment and responses to interaction. *Journal of Social and Personal Relationships, 18,* 263–289.

Johnson, D. J., & Rusbult, C. E. (1989). Resisting temptation: Devaluation of alternative partners as a means of maintaining commitment in close relationships. *Journal of Personality and Social Psychology, 57,* 967–980.

Karney, B. R., & Bradbury, T. N. (1995). The longitudinal course of marital quality and stability: A review of theory, method, and research. *Psychological Bulletin, 118,* 3–34.

Kashy, D. A., & Kenny, D. A. (2000). The analysis of data from dyads and groups. In H. T. Reis & C. M. Judd (Eds.), *Handbook of research methods in social psychology* (pp. 451–477). New York: Cambridge University Press.

Kelley, H. H. (1983). Love and commitment. In H. H. Kelley, E. Berscheid, A. Christensen, J. H. Harvey, T. L. Huston, G. Levinger, et al. (Eds.), *Close relationships* (pp. 265–314). San Francisco: Freeman.

Kelley, H. H., & Thibaut, J. W. (1978). *Interpersonal relations: A theory of interdependence.* New York: Wiley-Interscience.

Kenny, D. A., Kashy, D. A., & Cook, W. L. (2006). *Dyadic data analysis.* New York: Guilford Press.

Kobak, R. R., & Sceery, A. (1988). Attachment in late adolescence: Working models, affect regulation, and representations of self and others. *Child Development, 59,* 135–146.

Lazarus, R. J., & Folkman, S. (1984). *Stress, appraisal, and coping.* New York: Springer.

Menzies-Toman, D. A., & Lydon, J. E. (2005). Commitment-motivated benign appraisals of partner transgressions: Do they facilitate accommodation? *Journal of Social and Personal Relationships, 22,* 111–128.

Mikulincer, M. (1998). Attachment working models and the sense of trust: An exploration of interaction goals and affect regulation. *Journal of Personality and Social Psychology, 74,* 1209–1224.

Mikulincer, M., & Florian, V. (1998). The relationship between adult attachment styles and emotional and cognitive reactions to stressful events. In J. A. Simpson & W. S. Rholes (Eds.), *Attachment theory and close relationships* (pp. 143–165). New York: Guilford Press.

Mikulincer, M., Florian, V., & Weller, A. (1993). Attachment styles, coping strategies, and post-traumatic psychological distress: The impact of the Gulf War in Israel. *Journal of Personality and Social Psychology, 64,* 817–826.

Mikulincer, M., & Shaver, P. R. (2007). *Attachment in adulthood: Structure, dynamics, and change.* New York: Guilford Press.

Rholes, W. S., Simpson, J. A., Campbell, L., & Grich, J. (2001). Adult attachment and the transition to parenthood. *Journal of Personality and Social Psychology, 81,* 421–435.

Rusbult, C. E. (1983). A longitudinal test of the investment model: The development (and deterioration) of satisfaction and commitment in heterosexual involvements. *Journal of Personality and Social Psychology, 45,* 101–117.

Rusbult, C. E., Arriaga, X. B., & Agnew, C. R. (2002). Interdependence in close relationships. In G. J. O. Fletcher & M. S. Clark (Eds.), *Blackwell handbook of social psychology: Interpersonal processes* (pp. 359–387). Malden, MA: Blackwell.

Rusbult, C. E., Bissonnette, V. L., Arriaga, X. B., & Cox, C. L. (1998). Accommodation processes during the early years of marriage. In T. Bradbury (Ed.), *The developmental course of marital dysfunction* (pp. 74–113). New York: Cambridge University Press.

Rusbult, C. E., Verette, J., Whitney, G. A., Slovik, L. F., & Lipkus, I. (1991). Accommodation processes in close relationships: Theory and preliminary empirical evidence. *Journal of Personality and Social Psychology, 60*, 53–78.

Rusbult, C. E., Yovetich, N. A., & Verette, J. (1996). An interdependence analysis of accommodation processes. In G. J. O. Fletcher & J. Fitness (Eds.), *Knowledge structures in close relationships: A social psychological approach* (pp. 63–90). Mahwah, NJ: Lawrence Erlbaum Associates, Inc.

Rusbult, C. E., & Zembrodt, I. M. (1983). Responses to dissatisfaction in romantic involvements: A multidimensional scaling analysis. *Journal of Experimental Social Psychology, 19*, 274–293.

Shaver, P. R., & Hazan, C. (1993). Adult romantic attachment: Theory and evidence. In D. Perlman & W. Jones (Eds.), *Advances in personal relationships* (Vol. 4, pp. 29–70). London: Jessica Kingsley.

Simpson, J. A. (1990). Influence of attachment styles on romantic relationships. *Journal of Personality and Social Psychology, 59*, 971–980.

Simpson, J. A. (2007). Foundations of interpersonal trust. In A. W. Kruglanski & E. T. Higgins (Eds.), *Social psychology: Handbook of basic principles* (2nd ed., pp. 587–607). New York: Guilford.

Simpson, J. A., Rholes, W. S., & Phillips, D. (1996). Conflict in close relationships: An attachment perspective. *Journal of Personality and Social Psychology, 71*, 899–914.

Snyder, M., & Stukas, A. A. (1999). Interpersonal processes: The interplay of cognitive, motivational, and behavioral activities in social interaction. *Annual Review of Psychology, 50*, 273–303.

Thibaut, J., & Kelley, H. H. (1959). *The social psychology of groups*. New York: Wiley.

Tran, S., & Simpson, J. A. (2009). Pro-relationship maintenance behaviors: The joint roles of attachment and commitment. *Journal of Personality and Social Psychology, 97*, 685–698.

Van Lange, P., Rusbult, C. E., Drigotas, S. M., Arriaga, X. B., Witcher, B. S., & Cox, C. L. (1997). Willingness to sacrifice in close relationships. *Journal of Personality and Social Psychology, 72*, 1373–1395.

Vohs, K. D., & Finkel, E. J. (Eds.). (2006). *Self and relationships: Connecting intrapersonal and interpersonal processes*. New York: Guilford Press.

Wieselquist, J., Rusbult, C. E., Foster, C. A., & Agnew, C. R. (1999). Commitment, pro-relationship behavior, and trust in close relationships. *Journal of Personality and Social Psychology, 77*, 942–966.

CHAPTER 7

Identification

The Why of Relationship Commitment

JOHN E. LYDON

McGill University

LISA LINARDATOS

McGill University

Relationships are good for us—people with close relationships are happier and healthier than the lonely and socially isolated (Diener, Suh, Lucas, & Smith, 1999; House, Landis, & Umberson, 1988; Kiecolt-Glaser & Newton, 2001; Uchino, Cacioppo, & Kiecolt-Glaser, 1996). Theories of psychological needs consistently include relatedness as a basic human need (Baumeister & Leary, 1995; Deci & Ryan, 2000) and evolutionary theorists postulate why close relationships have had paramount importance for the survival and flourishing of the human species (e.g., Buss, 1994; Shaver, Hazan, & Bradshaw, 1988). Yet people often fail to sustain intimate relationships. The most obvious indicator of this is a high divorce rate, and we often overlook the fact that romantic relationships outside of marriage fail even more frequently than marriages.

Of course, there are a host of other things that are good for us that we fail at—healthy diets, regular exercise, daily flossing, to name a few. Unlike these, intimate relationships are typically fun, exciting, maybe even exhilarating experiences, particularly at the outset. That is, there is a strong approach motivation early on that fuels the development of the relationship, prompting the second and third dates and maybe even the 6-month anniversary. However, intimate relationships, like other long-term personal strivings, require sustained energy and effort.

Commitment represents the motivation to sustain energy and effort and to persist in the face of adversity. Consistent with a motivated cognition approach, commitment has been associated with a variety of cognitive and behavioral mechanisms that sustain relationships. Committed individuals attend less to attractive alternatives (Maner, Gailliot, & Miller, 2009; Miller, 1997), they make benign appraisals and benevolent attributions about the severity of partner transgressions (Finkel, Rusbult, Kumashiro, & Hannon, 2002; Menzies-Toman & Lydon, 2005), and they show a greater willingness to tolerate and forgive their partner's shortcomings (Rusbult, Verette, Whitney, Slovik, & Lipkus, 1991). Not surprisingly, commitment predicts long-term survival of intimate relationships (Le & Agnew, 2003), but it also predicts greater distress when a relationship fails (Lydon, Pierce, & O'Regan, 1997).

Whereas much research has examined how the degree of commitment promotes relationship survival, there has been scant attention given to the motivational bases of commitment and whether different motives differentially predict personal and relational well-being. A person may commit to a relationship because the relationship has been pleasant, enjoyable, and satisfying, what might be described as satisfaction-based commitment (M. Johnson, 1991) or enlightened self-interest. Some motivational theorists describe this as an intrinsic motivation to continue a relationship (Blais, Sabourin, Boucher, & Vallerand, 1990). A second reason to commit to a relationship is because it is seen as something one ought to do (Meyer & Allen, 1991). This has been referred to as introjection and represents an assimilation of values, standards, and pressures of others that induce guilt if not otherwise followed. A third reason, and the focus of our research, has been commitment because the relationship reflects a person's identity (Brickman, 1987; Burke & Reitzes, 1991), having been incorporated into one's representation of self. This is referred to as identification (Ryan, Rigby, & King, 1993).

It is likely that all three of these bases of commitment uniquely contribute to relationship functioning; however, we hypothesize that identification may be especially crucial in sustaining relationships in the face of adversity. Satisfaction-based commitment may be vulnerable to adversity because negative events are a direct challenge to the foundation of such commitment. If I am in the relationship because it is fun and enjoyable, then what do I do when it is not so fun and enjoyable? Ought-based commitments may help keep relationships intact out of a sense of duty and obligation but have a negative impact on well-being (Ryan & Connell, 1989), possibly by fostering resentment (Strauman & Higgins, 1988). Identification, in between introjection and intrinsic self-regulation on the self-determination continuum, may draw on some of the strengths of these other two motivations, without the corresponding weaknesses, to fuel commitment in a volitional manner.

Relationship Identification

There are at least two theoretical approaches from which to study identification: (1) a motivational perspective, as we have done in the academic domain (Burton, Lydon, D'Alessandro, & Koestner, 2006); and (2) a self-construal perspective (Cross, Bacon, & Morris, 2000). We review these theoretical approaches and demonstrate how we have applied them in our own research. Motivationally, identification has been described within a self-determination theory perspective as representing a degree of internalization of values and standards that may have originally been transmitted by others such as society, local community, or family but comes to reflect self-endorsed values and standards. Relationship researchers may think of this in terms of moral commitment (M. Johnson, 1991), but sometimes it is not clear whether a moral commitment reflects one's personally endorsed identification with the relationship or a weaker internalization of external standards that is reflected in guilt-based introjection (Pierce, Lydon, & Yang, 2001).

The concept of relationship identification follows from the work on relational schemas (Baldwin, 1992) and relational selves (Chen, Boucher, & Tapias, 2006), which illustrate the importance of how one mentally represents the relationship between self and other. Chen and her colleagues emphasize that the relational self is normative, consistent with the idea that relatedness is a fundamental psychological need. Because all of us need to negotiate relationships with others for survival and well-being, we develop mental representations of ourselves in relation to specific others (relationship-specific selves), and a summary representation of some of these relationship-specific selves may be abstracted in the form of a more generalized relational self (e.g., "Me when I am with my friends") as well as a global relational self that spans a wide range of relationships with significant others (e.g., "Me in relationships").

Chen and her colleagues review work on relational schemas, transference (Andersen & Chen, 2002), self–other representations (Ogilvie & Ashmore, 1991), and attachment (Mikulincer & Shaver, 2003) and propose that relational selves include both attribute and role-based conceptions and contain the affective cognitive units characteristic of cognitive affective personality systems theory (Mischel & Shoda, 1995), such as goals, motives, values, expectancies, and self-regulatory strategies. Thus, one can see that some of the theoretical features of identification from self-determination theory are central to a relational selves perspective. Moreover, relational selves include key features of relationship commitment—the motivation and procedural knowledge necessary to activate and apply relationship maintenance processes that help sustain relationships.

The normative focus of the relational self perspective is in contrast to individual differences in self-construal. Although everyone may have relational selves, there are likely individual differences in the extent to which the self in relation to close others is highly accessible and strongly linked with specific goals, values, and expectancies. We know, for example, that there are gender differences in the tendency to define oneself in terms of one's close relationships, with women having the tendency to incorporate close others into their self-conceptions more than men (Cross & Madson, 1997). In one clever illustration of this, female students given cameras and instructed to get 12 pictures that "represent you" were more likely to get pictures of themselves with other people than their male counterparts (Clancy & Dollinger, 1993).

Cross and Madson (1997) proposed that differences in self-construal should be associated with distinct patterns of motivation and cognition (Cross, Bacon, & Morris, 2000). These ideas spawned theory and research on the *relational-interdependent self-construal*, hereby referred to as the *relational self-construal.* Cross and colleagues have demonstrated that the Relational-Interdependent Self-Construal Scale (RISC Scale; Cross et al.), assessing individual differences in the degree to which individuals define themselves in terms of their close relationships, predicts a wide range of relationship phenomena. Evidence suggests, for example, that those with a highly relational self-construal have a rich and well-organized knowledge structure for information about one's close relationships (Cross, Morris, & Gore, 2002); self-disclose to their roommates more, which in turn promotes more positive evaluations from their roommates (Gore, Cross, & Morris, 2006); are more likely to consider the needs and wishes of others when making decisions (Cross et al.); and are more accurate in reporting on the beliefs and values of new roommates (Cross & Morris, 2003). It seems, because relationships are self-defining, individuals with a highly relational self-construal are motivated to protect and enhance close relationships.

Although we share the theoretical perspective of the relational self (Chen, Boucher, & Tapias, 2006) and its attention to the incorporation of relational experiences into self-representations and the social cognitive processes that ensue, our close relationships perspective emphasizes that relational experiences should influence individual differences in relational selves. Theory and research on the relational self-construal provide a platform for our exploration of relationship identification as a crucial ingredient in relationship commitment processes. We build on this and examine relationship-specific as well as more global relational self-construals, as people with similar global relational self-construals could differ in the extent to which they identify with their romantic relationship. Conversely, people with different global relational self-construals could be similar in the extent to which they identify with their romantic relationship. Moreover, and beyond the scope of our current work, individuals with a strong global relational self-construal may differ in the extent to which they identify with one close relationship compared with another close relationship.

Our empirical strategy was to start where Cross and colleagues started and illustrate gender differences in relationship maintenance processes. If such differences are obtained and reflect underlying differences in relational self-construal, then individual differences in the RISC Scale should predict relationship maintenance. Of particular interest to us was to then examine identification with a specific romantic relationship, which we refer to as *relationship-specific identification*. We discuss how we examined these individual differences in defining oneself in terms of one's current romantic relationship and then present findings on the correlates and consequences of having a strong relationship-specific identity.

Individual Differences in Relational Self-Construals

Given that women identify more with their relationships in general, we reasoned that women would be predisposed to exhibit behaviors that support and protect their relationships. However, because their identification may represent a generalized abstraction of their wide social network, they may require contextual cues to shift their attention from an abstract general identification to a concrete focus on one person in particular—their romantic partner. Highlighting the romantic relationship activates the specific relationship as an exemplar of the wider class of relationships that represent an individual's relational self. Moreover, it should also activate the affective and cognitive mediating units associated with a specific relationship identity, making the association between the self and the specific relationship prepotent if and when one subsequently encounters negative information about the relationship or the romantic partner. In other words, focusing on the romantic relationship provides a mental framework that would bias the processing and interpretation of subsequent information.

In the following study, we hypothesized that the positive goals, values, and expectancies associated with the relationship would be especially strong for women and could be activated prior to a negative encounter. We used a priming or salience manipulation to bring the romantic relationship into focal awareness before participants were asked to contemplate negative behaviors committed by their partners. One way to reduce the impact of a partner's negative behaviors is to make benevolent attributions for their transgressions. That is, one can attribute the causes to external, unstable, and specific factors rather than to stable, global, and internal factors. For example, Jack might attribute his partner's bad behavior to her being treated badly at work that day, an unusual circumstance that made his partner impatient for the first hour home from work. A person can also attribute responsibility in a benevolent way—by not blaming the partner or assuming selfish motives. We predicted that, when women were primed, they would increase their willingness to make benevolent attributions for their romantic partner's bad behavior, essentially being more forgiving. Men, who are less inclined to define themselves in terms of their dyadic relationships, should not be influenced by the prime. A prime presupposes a readiness to think in a particular way that in this case is less likely for men than women on average.

In one condition, the relationship prime was rather heavy-handed—completing a scale about relationship satisfaction and commitment (but keep in mind that relationship researchers do things like this in many studies without considering the possible effects on subsequent responses to later questions). We also created a second priming condition in which participants were simply asked a series of factual questions about their partner (e.g., age, how far apart they live, when they started dating). The dependent measure was the Relationship Attribution Measure (Fincham & Bradbury, 1992), whereby participants were asked to make attributions for prototypic negative partner behaviors. This measure consists of four negative, hypothetical partner behaviors (i.e., your partner criticizes something you say, your partner begins to spend less time with you, your partner does not pay attention to what you are saying, your partner is cool and distant), with six possible causal and responsibility attributions for each of the behaviors.

As expected, women primed with their relationship satisfaction and commitment made more benevolent attributions for their partner's negative behaviors (M = 3.09) than those in the no-prime control condition (M = 2.63). Importantly, women primed with relationship facts also made more benevolent attributions compared with those in the no-prime control condition (M = 3.12 vs. M = 2.63), F (2, 71) = 4.61, p = .01. Neither prime had an effect on men's attributions, $F > 1$. We assumed that men were not inclined to think of their network of dyadic relationships as self-defining, and therefore the prime proved ineffective.

Given the null results for men, we reasoned that the challenge was to get them thinking in more relational terms. Consequently, we did another experiment with the same dependent measure, but this time participants were primed with independence or interdependence using a word search task (Brewer & Gardner, 1996). In the independent prime condition, participants were asked to read a paragraph and circle the pronouns "it" and "its."

In the interdependent prime condition, these pronouns were replaced with the first-person plural pronouns "we," "us," and "our," which participants were asked to find and circle. Results revealed that men made more benevolent attributions when primed with "we" ($M = 3.48$) than when primed with "it" ($M = 2.92$). There was no effect for women. We assume that because women are already more inclined to think in terms of "we," the prime proved to be ineffectual. Whereas men need to shift from an independent self-construal to an interdependent self-construal to motivate relationship maintenance, women have numerous "we" relationships and the cognitive task is to focus on one in particular—the romantic relationship.

Chronic Relationship Identification

For us and others (e.g., Cross & Madson, 1997), gender differences often serve as a rough proxy for underlying psychological differences in relational self-construals. We reasoned that individual differences on the RISC Scale would be a more precise and reliable predictor of priming effects on relationship attributions compared with gender. We proceeded to develop a new relationship prime, whereby participants visualized being in the presence of their partner. To disguise the purpose of the study, participants were told that we were studying differences in visualizing social and nonsocial objects and were led to believe that they were randomly choosing one of six objects (three social and three nonsocial) to visualize. The control group members visualized their favorite food as a way to keep the valence of the visualization similar. Moreover, participants were asked questions not only about their relationship but also about their favorite food.

We obtained a significant prime by RISC Scale interaction effect ($p = .011$), such that individuals high on the RISC Scale made more benevolent attributions in the partner prime condition compared with each of the other three conditions, as revealed by a planned contrast, $t(136) = -2.75$, $SE = .12$, $p = .01$. Levels of relationship satisfaction and commitment were associated with benevolent attributions, but the prime by RISC Scale interaction remained significant controlling for satisfaction and for commitment, p's $< .05$. Thus, those with a highly relational self-construal were more forgiving of their partner's hypothetical negative behavior when reminded of their partner, regardless of their satisfaction with and commitment to the relationship.

These results (Linardatos & Lydon, 2011) reinforce and extend the previous findings on gender. In both cases, individual differences interacted with the prime attesting to the importance of individual differences in self-construal in promoting relationship maintenance processes. However, the findings also suggest that the power of such self-construals is at least in part a function of clear, salient linkages to the specific relationship. That is, relational self-construals may be more potent when the generalized, abstract self-construal is projected onto the specific relationship or relational partner prior to a negative interaction with one's partner because priming the self in relation to a significant other can activate a positive mental set about the relationship that motivates an individual to dampen the implications of a partner's negative behavior.

Identifying With a Specific Romantic Relationship

Of course, some individuals readily and easily construe their romantic relationship as a part of their identity without needing a prime. As mentioned already, we refer to the phenomenon of defining oneself in terms of one's specific relationship as relationship-specific identification. This concept reflects the idea that actual experiences in a dyadic relationship should have some influence on the degree to which one identifies with that specific relationship. We assume, as has been found with attachment (Pierce & Lydon, 2001), that identification operates both at the global and specific level. Next, we describe how relationship-specific identification may develop in a relationship and offer some tentative findings on the correlates of relationship-specific identification, which are in line with our hypotheses. We then discuss the utility of relationship-specific identification as a construct that provides increased explanatory power in accounting for relatively spontaneous and automatic relationship maintenance processes. But first, we start with findings demonstrating the convergent and discriminant validity of relationship-specific identification.

Although we expect identification with a specific romantic relationship to be distinct from dispositional relational self-construals, we assume that the two should be highly correlated, given that both measures tap into the same construct, just at different levels of analysis. The two measures are distinct in that relationship-specific identification captures relationship-specific experiences and evaluations. These relationship-specific factors should also be captured by other relationship constructs, the most obvious being relationship commitment. Therefore, we expect relationship constructs to be correlated with relationship-specific identification as well.

Our instrument was an adaptation of the RISC Scale (Cross et al., 2000), hereafter referred to as the Specific RISC Scale, whereby we presented the same statements but changed the context from relationships in general to the romantic relationship in particular. The RISC Scale and the Specific RISC Scale were correlated, as one would expect, but we were heartened by the fact that the correlation was moderately high at .57, suggesting that the two scales were not redundant. Moreover, the Specific RISC Scale correlated with relationship commitment, $r(289) = .46$, and when commitment and the global RISC Scale were simultaneously entered to predict relationship-specific identification, each uniquely contributed to its prediction, t's > 7, p's $< .001$. These findings suggest that the Specific RISC Scale is indeed a measure of identification like the RISC Scale but captures important unique variance at the relational level.

We then examined whether the Specific RISC Scale captures an aspect of commitment distinct from satisfaction. Both the Specific RISC Scale and satisfaction were entered in a regression analysis to predict commitment and each uniquely predicted commitment, t's > 6, p's $< .001$. This finding is consistent with our idea that there are qualitative differences in the underlying bases of commitment and that identification can be distinguished from satisfaction-based commitment.

To assess discriminant validity, the Specific RISC Scale was correlated with the "Big Five" personality dimensions of neuroticism, extraversion, openness to experience, conscientiousness, and agreeableness. None of the correlations were significant (all r's $< .10$). We also examined attachment avoidance and attachment anxiety (Collins & Read, 1990) in the context of relationships in general and found that these dimensions were not reliably correlated with identification (r's $< .20$). However, attachment avoidance in the context of a *specific* romantic relationship was negatively associated with identification ($r = -.44$). In other words, people that generally avoid closeness and dependency are not especially likely to refrain from identifying with their romantic relationship, but a lack of identification with a specific relationship corresponds to some degree with a fear of closeness and dependency about that specific relationship.

Bases of Relationship-Specific Identification

Although we have underscored a contrast between satisfaction-based intrinsic motivation and identification, we believe that pleasant and passionate experiences early in a relationship serve as key precursors to relationship-specific identification by fostering intrinsic motivation, which in turn should motivate increased frequency and diversity of interactions, thereby providing more opportunities for expressions of intimacy. In addition, with increased involvement, self-presentational concerns should wane, and individuals will increasingly risk revealing their hopes fears, goals, values, idiosyncrasies, and shortcomings, again promoting intimacy. Moreover, with increased interactions, there arise more conflicts of interest, providing opportunities to sacrifice self-interest on behalf of the partner, and sacrificing self-interest is a sign of intimacy.

We expect that intimacy will be an especially strong proximal predictor of relationship-specific identification. Individuals experience intimacy when they reveal their true self to their partner and feel understood, accepted, and cared for by their partner (Reis & Shaver, 1988). We theorize that, by revealing their true selves and receiving an intimate response from their partners, individuals will experience a stronger link between self and other, thereby identifying with the relationship. Moreover, the intimacy process is likely to influence the partner's identification because the partner now knows the person in a deeper, more intimate way that elicits communal behavior from the partner. Furthermore, Reis and Shaver speculate that, when individuals reveal their true selves, partners will be more likely to respond by revealing their true selves as well.

As a preliminary examination of the link between intimacy and relationship-specific identification, we used a person-perception paradigm to manipulate the presence of intimacy in a relationship and assess judgments of relationship-specific identification. Participants read one of four scenarios about Anne-Marie and Eric. We manipulated whether the scenario included words prototypical of intimacy in general or words prototypical of sexual intimacy. We also manipulated whether the other construct was not mentioned or was mentioned explicitly as lacking in the relationship. Participants then rated their impressions of the couple.

Participants perceived more relationship-specific identification when intimacy was present than when it was not present. We obtained the same pattern of results for perceptions of commitment, but the effect was not due to positive feelings about the relationship because we obtained a different pattern of results for relationship satisfaction. In this case, intimacy with an explicit lack of sexual intimacy led to lower perceptions of satisfaction than when sexual intimacy was present and intimacy was not mentioned. Thus, the lack of sexual intimacy signaled less satisfaction, but the presence of intimacy signaled greater identification. These results suggest that people associate intimacy particularly with relationship-specific identification and not just relationship positivity. Longitudinal or diary data may allow us to see if experiences of intimacy elevate relationship-specific identification.

Relationship-Specific Identification and Spontaneous Relationship Maintenance Responses

One reason for our interest in relationship-specific identification is our belief that it captures some of the essential variance in commitment that drives relationship maintenance responses. Typically, relationship maintenance is assessed with self-reports of an individual's willingness to accommodate and tolerate a partner's transgressions, to make sacrifices for the relationship, or to make benevolent attributions such as those we reported earlier. There is also a line of research that examines explicit judgments of the attractiveness of available alternative relationship partners.

We expected that relationship-specific identification should correlate with these sorts of measures (and it does), but we theorized that it would also be associated with even more spontaneous relationship maintenance responses. Specifically, because identification suggests an internalization of the relationship into one's core sense of self, it should be highly accessible and therefore able to influence relationship maintenance in a relatively effortless and maybe even automatic fashion.

We developed a new paradigm to test whether identification predicts spontaneous relationship-relevant behavior. To create a situation that would minimize the opportunity to provide deliberative responses, eliciting more spontaneous responses, we used Instant Messenger (IM), a popular online chat program. Heterosexual participants communicated over IM with an attractive confederate of the opposite sex, responding quickly to the questions posed by the confederate. We were interested in the frequency and immediacy with which they mentioned their dating partner. Our hypothesis was that relationship-specific identification would predict mentions of relationship status to the confederate.

The study was ostensibly about media, technology, and personality. Prior to the lab session, participants completed a premeasure of the Specific RISC Scale. Short introductions of the confederate were prerecorded and standardized and presented via a fake webcam to appear live. After some filler media tasks (watching news clips) and the short introductions, the next media task was the IM task. Participants were led to believe that they were randomly assigned to answer the questions of their IM partner, whereas their IM partner had ostensibly been randomly assigned to ask the questions. The questions were designed to first build rapport and to gradually become more personal and more likely to elicit mentions of one's dating partner (e.g. "Where are you from?" "What do you normally do on the weekends?" "Do you have any big plans for the summer?"). Frequency and immediacy with which participants mentioned their dating partner in the IM chat served as the dependent variables. We predicted that greater relationship-specific identification (as measured by the Specific RISC Scale prior to our lab session) would predict more immediate and frequent mentions of relationship status to the confederate.

In the first IM study, relationship-specific identification predicted number of mentions of the dating partner, $r(40) = .39$, $p = .01$. There are several possible explanations for this finding. It is possible that the highly identified participants were warding off the threat of an attractive alternative. Alternatively, because they define themselves in terms of their romantic relationship, they may be more inclined to disclose their relationship status to people in general. To test these possibilities, we conducted a second study in which we manipulated whether heterosexual participants interacted with an attractive opposite-sex confederate or an attractive same-sex confederate.

In IM Study 2, we obtained the same significant correlation between relationship-specific identification and the frequency of partner mentions in the opposite-sex condition, $r(54) = .485$, $p < .0001$, but this was not found in the same-sex condition, $r(61) = .09$. In addition, the condition by Specific RISC Scale interaction predicted the frequency with which the partner was mentioned ($B = .181$, $SE = .090$, ß $= .178$, $t = 2.02$, $p = .046$). A closer inspection of the means revealed that low identifiers mentioned their partner less than high identifiers in the opposite-sex condition ($B = .446$, $SE = .124$, $t = 3.60$, $p < .0001$) and less than low identifiers in the same-sex condition ($B = -.247$, $SE = .126$, $t = -1.95$, $p = .053$), whereas high identifiers did not differ in their number of partner mentions between conditions ($t < 1$). These effects held controlling for the dispositional measure of relational self-construal and for relationship commitment.

We obtained the same significant interaction when we examined the number of questions that needed to be asked to elicit mention of the partner and when we examined the likelihood of never mentioning the partner at all. Whereas most individuals mentioned their partner by the tenth question ("What was the last really fun thing you did?"), a significantly greater percentage of those who were less identified with their relationship refrained from mentioning their partner to the attractive opposite-sex confederate at this point. In fact, less than half of this group ever mentioned their partner at all, whereas almost 90% of those who were highly identified with their relationship mentioned their partner to the opposite-sex confederate at least once.

Although the demands of the IM paradigm require rather quick, spontaneous responses, they do not require the extremely fast responses characteristic of social cognitive reaction-time measures. Thus, to push our tests of relationship-specific identification further, we conducted a study of participants' attention to attractive alternatives using a dot-probe paradigm. In this paradigm heterosexual participants are presented with distractor images in one of four quadrants on a computer screen, followed by a circle or a square that they are asked to categorize (Maner, Gailliot, & DeWall, 2007). In our study, the distractors were images of attractive and average-looking men and women. Attentional adhesion represents the degree to which one's attention is captured by and stuck on the distractor image (e.g., attractive opposite-sex image), thereby slowing the shifting of attention to another quadrant to categorize the response object as a circle or a square.

The simple prediction would be that relationship-specific identification would be associated with attentional adhesion, such that high identifiers would show less attentional adhesion to the attractive opposite-sex image relative to the other three sets of images. However, Lydon, Menzies-Toman, Burton, and Bell (2008) proposed that this sort of automatic attentional effect might be triggered by a relevant relationship threat—the anticipation of an interaction with an attractive available person of the preferred sex. Therefore, we manipulated this by presenting introductory videos of attractive male and female confederates prior to the dot-probe task and leading participants to believe that they would have an interaction with the confederate later in the experimental session.

The results revealed a marginal three-way interaction among relationship-specific identification, threat (anticipated interaction with attractive alternative), and image (attractive opposite-sex image vs. other images; $F(1, 72) = 2.95, p < .10$). Simple effects tests revealed that in the threat condition low identifiers showed more attentional adhesion to the attractive alternative compared with the other images whereas high identifiers showed less attentional adhesion to the attractive alternative compared with the other three images ($p < .01$).

Collectively, these results indicate that relationship-specific identification is likely an important ingredient in the prediction of relationship maintenance processes. Its ability to predict relatively spontaneous responses is especially important given the likelihood that, in the real world, people often are called upon and challenged to exhibit relationship maintenance responses when their self-regulatory resources are low (e.g., due to fatigue) or when the demands of the situation require a very rapid and unconsidered response.

Motivation and Daily Experiences

Recognizing the importance of assessing real-world experiences, we sought to examine how identification might contribute to personal and relationship well-being when one is faced with negative relationship events or experiences a lack of positive relationship events. To examine this, we drew on the distinctions in motivation delineated by self-determination theory, as described in the introduction. Presumably, intrinsic motivation, which is based in enjoyable and satisfying experiences, will contribute to relationship regulation. We wanted to see if identification might contribute to relationship regulation in ways that are distinct from intrinsic motivation. Former doctoral student Danielle Menzies-Toman, along with Lydon (2010) examined the potential differences in daily experiences as a function of intrinsic and identified relationship motives. Drawing on previous measures of relationship motivation (e.g., Blais et al., 1990; Rempel, Holmes, & Zanna, 1985), we developed measures of intrinsic (five items), identified (seven items), and introjected (seven items) relationship motives. The prototypical intrinsic item was, "I am in my relationship because it brings me joy to be in my relationship"; the identified item was, "I am in the relationship because my relationship feels right to me"; and the introjected item was, "I am in my relationship because it is something I feel I should do."

We surveyed 621 individuals in romantic relationships to verify the internal reliability of each measure and to compare a three-factor solution to the measures of motivation versus one- and two-factor solutions. All three measures had high Cronbach's alphas (.83–.89). Confirmatory factor analysis revealed that a three-factor solution provided a better fit for the data than one general factor of motivation, a two-factor solution comparing introjected motives to a factor of intrinsic and identified motives, and a two-factor solution comparing intrinsic motives to a factor of identified and introjected motives.

We then proceeded to conduct a study of couples and their daily experiences. A total of 63 couples came to the lab and completed a battery of measures, including the measures of relationship motives, and were given instructions as to how to report their daily interactions. They were told to report on every distinct positive and negative interaction with their partner, writing a very brief factual description of the event and rating its positivity and negativity, and to ignore interactions that fell within the neutral zone of affectivity. Additionally, in the evenings, participants were instructed to complete a battery of questions on their personal and relationship well-being.

The majority of daily events recorded were positive (78%), and, of these, 53% were reported as partner initiated, 28% as initiated by both members, and 20% as self-initiated. All events were read and rated by two members of the research team in terms of the severity of negatives and the positivity of positives. Motives were found to be associated with objective ratings by raters. In particular, individuals high in intrinsic motivation were in relationships characterized as having less objectively severe negative events, and individuals high in identified motivation were in relationships in which the positive events were seen as objectively more positive. Additionally, identified motives were associated with reporting more positives performed by the partner than the partner reported doing. Although correlational, these results suggest that objective daily experiences may be reinforcing these relationship motives or that the motives elicit more favorable behavior from the partner, or both.

The primary questions we had were whether (1) the frequency of positive or negative events predicts changes in personal or relational well-being, and (2) whether motives moderate such associations. Because daily experiences of positivity and negativity on one hand and well-being on the other were nested within individuals who varied in terms of relationship motives, the data were analyzed using hierarchical linear modeling.

Results revealed that the frequency of negative events was associated both with decreases in relationship well-being and personal well-being

(p's < .01). However, the motivational basis for the relationship interacted with the frequency of negative events to predict each type of well-being. As one might expect, the more that a person was in the relationship for the fun and enjoyment of it, the more that negative events decreased their relationship happiness ($p = .01$). However, this was not the case for identified motives. Instead, identified motives interacted with negative events to predict a decrease in personal happiness ($p = .02$). That is, the more that the relationship is an important, meaningful representation of who I am, the more that negative relationship events are associated with a decline in how I feel about my life.

Whereas the frequency of positive events did not predict changes in well-being during the 2-week period, it did predict changes in commitment. Consistent with an enlightened self-interest approach to commitment, a low frequency of positive events predicted a drop in relationship commitment. Based on the assumption that identification should represent an aspect of commitment that is relatively less sensitive to the hedonic value of immediate positive events, we hypothesized that identified motives would moderate the positive events to commitment association. Indeed, positive events predicted changes in commitment, but this depended on the level of identified motives. The interaction was due to people who were low in identified motives reporting less commitment at the end of the 2 weeks if they had relatively few positive interactions during the 2 weeks. Individuals high in identified motives were immune to this—they remained equally committed, even if there had not been many positive interactions during the 2-week period.

A subsample of participants from the original scale validation survey of relationship motives in Menzies-Toman and Lydon (2010) were recontacted to ascertain whose relationships had survived. Of the 141 contacted, 32 participants reported that their relationships had ended. Relationship persistence (broken up versus still together) after 6 months was examined using a logistic regression analysis, and changes in relationship satisfaction were examined using a linear regression analysis in which time 1 satisfaction was a baseline predictor variable and time 2 satisfaction was the criterion.

Although intrinsic ($\beta = .08$, $t(140) = 5.06$, $p < .05$) and identified ($\beta = .08$, $t(140) = 9.60$, $p < .05$) motives predicted staying together at time 2, when both predictors were entered simultaneously in the regression analysis, identified motives emerged as a significant predictor of staying together ($\beta = .10$, $t(137) = 5.00$, $p < .05$), whereas intrinsic motives no longer accounted for why couples stayed together ($\beta = -.03$, $t(137) = .23$). We conducted this analysis again using intrinsic and identified motive measures split on the median to assess the odds ratio of staying together as a function of identified motives. This analysis resulted in an odds ratio of

2.25 for individuals high in identified motives. That is, individuals high in identified motives were more than twice as likely to remain together 6 months later.

Because identified motives both predicted relationship persistence and correlated with commitment, and commitment was also a predictor of persistence at time 2, the preconditions for mediation (Independent Variable → Dependent Variable, IV → Mediating Variable, and MV → DV) were met, and mediation was formally assessed using the Sobel test. Commitment predicted relationship persistence ($\beta = .08$, $t(143) = 17.13$, $p < .01$), and commitment mediated the influence of identified motives on persistence (Sobel's $z(142) = 3.35$, $p < .01$).

Introjected motives also predicted staying together at time 2 ($\beta = .06$, $t(140) = 5.58$, $p < .05$). We conducted this analysis again using the introjected motives measure split on the median to assess the odds ratio of staying together as a function of introjected motives. This analysis resulted in an odds ratio of 2.64 for individuals high in introjected motives. That is, individuals high in introjected motives were almost three times as likely to remain together 6 months later. However, introjected motives predicted declines in time 2 satisfaction (controlling for time 1 satisfaction; $\beta = -.19$, $t(86) = -2.49$, $p = .02$) for those who remained together. Thus, as hypothesized, although those who were introjected were more likely to stay together, they were less satisfied at time 2. In contrast, those high in identified motives were marginally more satisfied at reassessment, as identified motives predicted time 2 satisfaction (controlling for time 1 satisfaction; $\beta = .18$, $t(88) = 1.92$, $p = .06$) for those who remained together.

The longevity results are consistent with work using the Specific RISC Scale, not surprisingly as identified motives and relationship-specific identification are highly correlated ($r = .71$). Linardatos and Lydon (2011) recontacted 283 participants from studies in which we had administered the Specific RISC Scale and assessed relationship status 1 to 3 years after the initial assessment. A total of 191 participants reported that they were still in the same relationship, and 92 reported that they were no longer in that relationship. A logistic regression analysis revealed that relationship-specific identification predicted relationship survival ($\chi^2 = 10.04$, $p < .01$) with an odds ratio of 1.50. Thus, individuals high on the Specific RISC Scale were 50% more likely to be in the same relationship. This is especially remarkable considering it takes a lack of commitment on the part of only one member of the dyad to end a relationship, and we had data on only one of the members.

A subset of these participants also completed measures of the RISC Scale, used to assess relational self-construal, and relationship commitment. Relationship-specific identification predicted relationship survival when controlling for the RISC Scale (OR = 2.00, $p = .03$) but not when controlling for commitment ($p = .23$), whereas commitment did predict relationship survival (OR = 1.85, $p < .01$). Thus, relationship-specific identification is a better predictor of relationship survival when compared with one's tendency to identify with relationships in general, but commitment is a better predictor of relationship survival when compared to relationship-specific identification.

Relationship-Specific Identification and Commitment

Theoretically, we might expect the effects of relationship-specific identification to be mediated by commitment, and occasionally we obtain such results. However, in some studies, particularly studies of proximal relationship maintenance responses, such as those that used the IM paradigm, identification often outperforms commitment in predicting the response. We suspect that this is the case because identification captures a particular kind of commitment that is crucially important in relationship maintenance situations. That is, because identification includes the social cognitive architecture of relational selves, persons who are highly identified with their relationship should have a highly accessible and elaborated representation of self in relation to other, which would include procedural knowledge that can be readily called upon as needed.

In measurement terms, commitment likely has greater bandwidth because it reflects relationship satisfaction and obligation (introjection) as well as identification. If these are not as directly tied to relationship maintenance as identification-based commitment, then a measure that includes them may not perform as well. However, over the long course of a relationship, we expect that satisfaction helps to fuel commitment, and, as indicated previously, introjection is a good predictor of relationship survival. Thus, a measure of relationship commitment that incorporates all of these factors should be an especially robust predictor of relationship survival.

Dark Side of Identification

Although throughout this chapter we have focused on how identification may motivate people to engage in thoughts and behaviors to help sustain their relationships, this may come with costs. First, relationship maintenance is agnostic about whether maintaining the relationship is good for one's personal well-being. People can tolerate unhealthy and even abusive relationships (Arriaga, 2002). It is somewhat encouraging that, at least in our experience sampling data, identification was associated with personal well-being in the face of negative events, but this was just within a 2-week period and for students in nonmarital relationships. It is quite possible that over time identification may lead people to tolerate increasingly negative partner behaviors.

Another way identification may be detrimental is in adjustment to relationship dissolution. We have been investigating the clinging to an apparently unsuccessful relationship in terms of what we call "lingering identification." In our longitudinal study of relationship-specific identification, we adapted the measure to assess lingering identification in individuals whose relationships had ended. We were interested in the correlates of lingering identification—What are the negative effects associated with hanging onto an old relationship?

We (Linardatos & Lydon, 2009) recontacted 92 participants whose relationships had ended and had completed the Specific RISC Scale 1 to 3 years earlier while in a dating relationship. At this postbreakup, time 2 period, we assessed lingering identification by having participants rate their agreement with statements such as, "My past romantic relationship is an important reflection of who I am." We also asked participants about the degree to which their relationship breakup interfered with personal activities the first 2 months postbreakup and currently interfered with activities such as school and work. Participants also reported on their degree of rumination by responding to statements such as, "These days, how often do you become absorbed or 'caught up' in thoughts and memories of your former relationship or former partner?" (Gray & Silver, 1990). Finally, we had participants rate their subjective distance from their prebreakup self.

Time 1 relationship-specific identification correlated with time 2 lingering identification ($r(83) = .36$, $p < .01$). Time 1 identification also correlated with reports of postbreakup interference with activities ($r(83) = .28$, $p < .05$), and this was mediated by lingering identification, suggesting that the difficulty of disengaging from a previous relational identity may interfere with daily functioning. This was further supported by the correlation between lingering identification and current interference with activities ($r = .36$, $p < .01$) and between lingering identification and rumination ($r = .53$, $p < .01$).

Whereas the more proximal measure of lingering identification generally outperformed the more distal measure of relationship-specific identification, there was one exception. Time 1 identification predicted subjective distance ($r = .28$, $p < .05$), whereas lingering identification did not ($r = -.06$). Supplementary analyses suggest that people who identified with their romantic relationship may have experienced two postbreakup paths. Some remained identified and could not disengage. As a result, the former relationship had a persistent and pernicious effect on other aspects of their lives. In contrast, others subjectively distanced themselves and became less preoccupied with the past relationship, thereby dampening rumination and interference with personal activities.

Conclusion

William James wrote that we have many selves but that we must review the list and choose the one on which we "stake our salvation." He expanded on this idea to present self-esteem as a function of success of important selves. The implication is that if I do not identify with something, then it will not influence my self-esteem. Positive relationship experiences can make a person feel good, but the true power of a relationship to influence, and maybe even transform, one's self-worth is contingent on one taking the risk to identify with the relationship, to staking a claim that the relationship is indeed an important, integral part of one's self. It may be safer to hold back and not identify, thereby avoiding the potential pain of rejection or betrayal. But a lack of identification may weaken one's ability to think and do the things that help sustain the relationship, and a self-protective reticence to identify may fuel a self-fulfilling prophecy. Besides, according to Lord Alfred Tennyson, "It is better to have loved and lost than never to have loved at all."

References

Agnew, C., Van Lange, P., Rusbult, C., & Langston, C. (1998). Cognitive interdependence: Commitment and the mental representation of close relationships. *Journal of Personality and Social Psychology, 74,* 939–954.

Andersen, S. M., & Chen, S. (2002). The relational self: An interpersonal social-cognitive theory. *Psychological Review, 109,* 619–645.

Arriaga, X. B. (2002). Joking violence among highly committed individuals. *Journal of Interpersonal Violence, 17,* 591–610.

Baldwin, M. W. (1992). Relational schemas and the processing of social information. *Psychological Bulletin, 112,* 461–484.

Baumeister, R. F., & Leary, M. R. (1995). The need to belong: Desire for interpersonal attachments as a fundamental human motivation. *Psychological Bulletin, 117,* 497–529.

Blais, M. R., Sabourin, S., Boucher, C., & Vallerand, R. J. (1990). Toward a motivational model of couple happiness. *Journal of Personality and Social Psychology, 59,* 1021–1031.

Brewer, M. B., & Gardner, W. (1996). Who is this "we"? Levels of collective identity and self-representations. *Journal of Personality and Social Psychology, 71,* 83–93.

Brickman, P. (1987). Commitment. In B. Wortman & R. Sorrentino (Eds.), *Commitment, conflict, and caring* (pp. 1–18). Englewood Cliffs, NJ: Prentice Hall.

Burke, P. J., & Reitzes, D. C. (1991). An identity theory approach to commitment. *Social Psychology Quarterly, 54,* 239–251.

Burton, K. D., Lydon, J. E., D'Alessandro, D. U., & Koestner, R. (2006). The differential effects of intrinsic and identified motivation on well-being and performance: Prospective, experimental, and implicit approaches to self-determination theory. *Journal of Personality and Social Psychology, 91,* 750–762.

Buss, D. M. (1994). *The evolution of desire: Strategies of human mating.* New York: Basic Books.

Chen, S., Boucher, H. C., & Tapias, M. P. (2006). The relational self revealed: Integrative conceptualization and implications for interpersonal life. *Psychological Bulletin, 132,* 151–179.

Clancy, S. M., & Dollinger, S. J. (1993). Photographic depictions of the self: Gender and age differences in social connectedness. *Sex Roles, 29,* 477–495.

Collins, N. L., & Read, S. J. (1990). Adult attachment, working models, and relationship quality in dating couples. *Journal of Personality and Social Psychology, 58,* 644–663.

Cross, S. E., Bacon, P., & Morris, M. (2000). The relational-interdependent self-construal and relationships. *Journal of Personality and Social Psychology, 78,* 791–808.

Cross, S. E., & Madson, L. (1997). Models of the self: Self-construals and gender. *Psychological Bulletin, 122,* 5–37.

Cross, S. E., & Morris, M. L. (2003). Getting to know you: The relational self-construal, relational cognition, and well-being. *Personality and Social Psychology Bulletin, 29,* 512–523.

Cross, S., Morris, M., & Gore, J. (2002). Thinking about oneself and others: The relational-interdependent self-construal and social cognition. *Journal of Personality and Social Psychology, 82,* 399–418.

Deci, E. L., & Ryan, R. M. (2000). The "what" and "why" of goal pursuits: Human needs and the self-determination of behavior. *Psychological Inquiry, 11,* 227–268.

Diener, E., Suh, E. M., Lucas, R. E., & Smith, H. L. (1999). Subjective well-being: Three decades of progress. *Psychological Bulletin, 125,* 276–302.

Downey, G., & Feldman, S. (1996). Implications of rejection sensitivity for intimate relationships. *Journal of Personality and Social Psychology, 70,* 1327–1343.

Fincham, F. D., & Bradbury, T. N. (1992). Assessing attributions in marriage: The Relationship Attribution Measure. *Journal of Personality and Social Psychology, 62,* 457–468.

Finkel, E., Rusbult, C., Kumashiro, M., & Hannon, P. (2002). Dealing with betrayal in close relationships: Does commitment promote forgiveness? *Journal of Personality and Social Psychology, 82,* 956–974.

Gore, J. S., Cross, S. E., & Morris, M. L. (2006). Let's be friends: The relational self construal and the development of intimacy. *Personal Relationships, 13,* 83–102.

Gray, J. D., & Silver, R. C. (1990). Opposite sides of the same coin: Former spouses' divergent perspectives in coping with their divorce. *Journal of Personality and Social Psychology, 59,* 1180–1191.

House, J. S., Landis, K. R., & Umberson, D. (1988). Social relationships and health. *Science, 241,* 540–545.

Johnson, M. P. (1991). Commitment to personal relationships. In W. H. Jones & D. W. Perlman (Eds.), *Advances in personal relationships* (Vol. 3, pp. 117–143). London: Jessica Kingsley.

Kiecolt-Glaser, J. K., & Newton, T. L. (2001). Marriage and health: His and hers. *Psychological Bulletin, 127,* 472–503.

Le, B., & Agnew, C. R. (2003). Commitment and its theorized determinants: A meta-analysis of the investment model. *Personal Relationships, 10,* 37–57.

Linardatos, L., & Lydon, J. E. (2009, February). *Relationship identification over time: How continued identification with a former partner hinders personal and relational well-being.* Poster session presentation at the annual meeting of the Society for Personality and Social Psychology, Tampa, FL.

Linardatos L., & Lydon, J. E. (2011). A little reminder is all it takes: The effects of priming and relational self-construal on responses to partner transgressions. *Self and Identity, 10,* 85–100.

Linardatos, L., & Lydon, J. E. (2011, July 4). Relationship-specific identification and spontaneous relationship maintenance processes. *Journal of Personality and Social Psychology.* Advanced online publication retrieved from doi:10.1037/a0023647

Lydon, J., Menzies-Toman, D. A., Burton, K., & Bell, C. (2008). If-then contingencies and the differential effects of the availability of an attractive alternative on relationship maintenance for men and women. *Journal of Personality and Social Psychology, 95,* 50–65.

Lydon, J. E., Pierce, T., & O'Regan, S. (1997). Coping with moral commitment to long-distance dating relationships. *Journal of Personality and Social Psychology, 73,* 104–113.

Maner, J. K., Gailliot, M. T., & DeWall, C. N. (2007). Adaptive attentional attunement: Evidence for mating-related perceptual bias. *Evolution and Human Behavior, 28,* 28–36.

Maner, J. K., Gailliot, M. T., & Miller, S. L. (2009). The implicit cognition of relationship maintenance: Inattention to attractive alternatives. *Journal of Experimental Social Psychology, 45,* 174–179.

Markus, H., & Kitayama, S. (1991). Culture and the self: Implications for cognition, emotion, and motivation. *Psychological Review, 98,* 224–253.

Menzies-Toman, D. A., & Lydon, J. E. (2005). Commitment-motivated benign appraisals of partner transgressions: Do they facilitate accommodation? *Journal of Social and Personal Relationships, 22,* 111–128.

Menzies-Toman, D., & Lydon, J E (2010). *Self-determination theory and close relationships: Intrinsic, identified and introjected relationship motives.* Unpublished manuscript, McGill University.

Meyer, J. P., & Allen, N. J. (1991). A three-component conceptualization of organizational commitment. *Human Resource Management Review, 1,* 61–98.

Mikulincer, M., & Shaver, P. R. (2003). The attachment behavioral system in adulthood: Activation, psychodynamics, and interpersonal processes. *Advances in Experimental Social Psychology, 35,* 53–152.

Miller, R. (1997). Inattentive and contented: Relationship commitment and attention to alternatives. *Journal of Personality and Social Psychology, 73,* 758–766.

Mischel, W., & Shoda, Y. (1995). A cognitive-affective system theory of personality: Reconceptualizing situations, dispositions, dynamics, and invariance in personality structure. *Psychological Review, 102,* 246–268.

Ogilvie, D. M., & Ashmore, R. D. (1991). Self-with-other representation as a unit of analysis in self-concept research. In R. C. Curtis (Ed.), *The relational self: Theoretical convergences in psychoanalysis and social psychology* (pp. 282–314). New York: Guilford.

Pierce, T., & Lydon, J. (2001). Global and specific relational models in the experience of social interactions. *Journal of Personality and Social Psychology, 80,* 613–631.

Pierce, T., Lydon, J., & Yang, S. (2001). Enthusiasm and moral commitment: What sustains family caregivers of those with dementia. *Basic and Applied Social Psychology, 23,* 29–41.

Reis, H. T., & Shaver, P. (1988). Intimacy as an interpersonal process. In S. Duck (Ed.), *Handbook of personal relationships* (pp. 367–389). Chichester, England: Wiley.

Rempel, J., Holmes, J., & Zanna, M. (1985). Trust in close relationships. *Journal of Personality and Social Psychology, 49,* 95–112.

Rusbult, C. E., Verette, J., Whitney, G. A., Slovik, L. F., & Lipkus, I. (1991). Accommodation processes in close relationships: Theory and preliminary empirical evidence. *Journal of Personality and Social Psychology, 60,* 53–78.

Ryan, R. M., & Connell, J. P. (1989). Perceived locus of causality and internalization: Examining reasons for acting in two domains. *Journal of Personality and Social Psychology, 57,* 749–761.

Ryan, R. M., Rigby, S., & King, K. (1993). Two types of religious internalization and their relations to religious orientations and mental health. *Journal of Personality and Social Psychology, 65,* 586–596.

Sanderson, C. A., & Cantor, N. (1995). Social dating goals in late adolescence: Implications for safer sexual activity. *Journal of Personality and Social Psychology, 68,* 1121–1135.

Shaver, P., Hazan, C., & Bradshaw, D. (1988). Love as attachment: The integration of three behavioral systems. In R. J. Steinberg & M. L. Barnes (Eds.), *The psychology of love* (pp. 68–99). New Haven, CT: Yale University Press.

Sheldon, K., & Elliot, A. (1998). Not all personal goals are personal: Comparing autonomous and controlled reasons for goals as predictors of effort and attainment. *Personality and Social Psychology Bulletin, 24,* 546–557.

Strauman, T. J., & Higgins, E. T. (1988). Self-discrepancies as predictors of vulnerability to distinct syndromes of chronic emotional distress. *Journal of Personality, 56,* 685–707.

Uchino, B. N., Cacioppo, J. T., & Kiecolt-Glaser, J. K. (1996). The relationship between social support and physiological processes: A review with emphasis on underlying mechanisms and implications for health. *Psychological Bulletin, 119,* 488–531.

CHAPTER 8

Motivating Commitment

The Power of a Smart Relationship Unconscious

SANDRA L. MURRAY
University at Buffalo, State University of New York

REBECCA T. PINKUS
University of Western Sydney

> The uttered part of a man's life, let us always repeat, bears to the unuttered, unconscious part a small unknown proportion. He himself never knows it, much less do others.
>
> **Thomas Carlyle (1795–1881)**

While Thomas Carlyle probably did not intend his words as a commentary on romantic love, his insights have telling implications for close relationships nonetheless. Perhaps the "unuttered" or unconscious mind knows something about one's relationship that the "uttered" or conscious mind does not fully appreciate. Imagine the potentially unhappy relationship fate of a low-self-esteem Gayle. Her conscious mind makes it hard for her to sustain her commitment. She constantly worries that Ron does not love her as much as she loves him (Murray, Holmes, & Griffin, 2000). Such doubts typically weaken her commitment by motivating her to distance herself from Ron—rejecting him before he has the chance to reject her (Murray, Holmes, & Collins, 2006). Might Gayle's unconscious mind better know how to sustain her commitment?

Current social psychological theorizing and research on the "smart" unconscious give reason for optimism (Bargh, 2007; Bargh & Morsella, 2008; Dijksterhuis & Nordgren, 2006; Murray, Holmes, & Pinkus, 2010; Wegner, 2002; Wilson, 2002). In proposing a new model of interdependence, Murray and Holmes (2009, 2011) attributed part of the "know-how" to build stable relationship bonds to the unconscious. In this spirit, the current chapter reviews two lines of research that suggest that a smart relationship unconscious can sometimes better protect one's commitment than one's conscious mind. Through both lines of research, we argue that unconscious or automatic processes of partner valuing can help sustain relationship commitment in the face of sometimes disappointing realities, unless the conscious mind gets in the way.

This chapter unfolds in three major parts. In the introduction, we detail the model of the interdependent mind that provides the conceptual basis for our arguments. The term *interdependent mind* refers to an interconnected system of "if–then" rules for interaction within adult romantic relationships (Murray & Holmes, 2009, 2011). These unconscious rules coordinate partner interaction by linking specific features of the situation (i.e., "If Ron does X") to correspondent ways of feeling, thinking, and behaving (i.e., "Then Gayle does Y"). In the main body of the chapter, we describe how unconscious processes of partner valuing that result from applying these rules help sustain commitment in the face of costs. We highlight two lines of research. The first explores how Gayle's unconscious sustains her commitment in the face of the costs to her personal goals that naturally arise in living with Ron—from having her sleep disrupted by his snoring to enduring his inexplicable musical tastes (Murray, Holmes, et al., 2009). The second explores how Gayle's unconscious protects her commitment in the face of the hurts and slights that come from Ron's occasional, but inevitable, rejecting behaviors—from breaking his promise to do his household chores to listening half-heartedly to her work woes (Murray, Pinkus, Holmes, Harris, Gomillion, Aloni, et al., 2010). We conclude the chapter with reflections on how modeling the contents of the unconscious and conscious mind might deepen our understanding of close relationships.

The Interdependent Mind: A Smart Unconscious for Relationships

Adult romantic relationships offer the chance to satisfy basic human needs for social connection (Baumeister & Leary, 1995). Relationships realize this promise when partners coordinate their behavior to be consistently caring and responsive to one another's needs (Reis, Clark, & Holmes, 2004). Unfortunately, interdependence does not make it easy to be responsive (Kelley, 1979). Consider Gayle and Ron in this respect. She's a neat freak and he eschews order; he's a couch potato and she's an exercise enthusiast; he's an extravert and she's an introvert; she has traditional family values and he aspires to be a stay-at-home-dad. Such incompatibilities are more common than not in most relationships because partners do not selectively assort on basic dimensions of preference or personality (Lykken & Tellegen, 1993). Consequently, partners invariably encounter situations where their personal interests or preferences conflict (Murray & Holmes, 2009). Such conflict-of-interest situations make it difficult for partners to be responsive to one another's needs because behavioral coordination involves sacrificing one's own self-interest. In his marriage, Ron cannot simultaneously meet Gayle's desire for quiet time with him without giving up some of his desire for social time with friends and family. Given the adaptive importance of the need to belong and the objective difficulty of behavioral coordination, it seems reasonable to assume that the social mind evolved mechanisms for promoting stable social bonds (Enfield & Levinson, 2006).

Murray and Holmes (2009, 2011) proposed the existence of an "interdependent mind" that functions to facilitate mutual responsiveness (see Bargh & Huang, 2009; Bargh & Morsella, 2008). This "smart" relationship unconscious motivates responsiveness through a system of interconnected if–then or procedural rules that help partners coordinate reciprocal expressions of commitment.[1] In colloquial terms, these rules generally match what Gayle asks of Ron to what Ron is willing to provide and match what Ron asks of Gayle to what Gayle is willing to provide. In formal terms, these rules link the risk properties of situations to correspondent interpersonal goals and behavioral strategies for goal pursuit (Murray, Aloni, et al., 2009; Murray, Holmes, et al., 2009; Murray, Derrick, Leder, & Holmes, 2008). Through these rules, the interdependent mind implicitly recognizes that promoting mutuality in responsiveness requires different tactics in low-risk situations where partner interests are relatively compatible than in high-risk situations where partner interests are relatively incompatible.

Coordinating Expressions of Commitment

Because situations vary in the level of risk they pose, the interdependent mind has two general strategies at hand for coordinating mutual responsiveness. These strategies work by matching one partner's expression of commitment to the other partner's expression of commitment in a way

[1] The term *reciprocal expressions of commitment* refers to equality in partners' expressions of commitment across time and situations within the relationship. Reciprocity does not require equal expressions of commitment in a specific situation within the relationship. It does require that expressions of commitment generally balance out across partners over successive interactions (Holmes, 1981; Reis, Clark, & Holmes, 2004).

that accommodates to situational risk. Before detailing these strategies, we first explain why reciprocated commitments are central to responsiveness.

Interdependence theorists define commitment in terms of the objective state of dependence and the subjective state of attachment (Rusbult & Van Lange, 2003). *Dependence* increases the more partners rely on one another to fulfill their needs (e.g., companionship, recreation, sex), the more time, energy, and resources they invest into the relationship, and the fewer alternatives they possess to being in the relationship. This means that partners can express commitment through the concrete ways they put their outcomes in one another's hands and tend to one another's outcomes. For instance, Gayle can express her commitment to Ron when she relies on him for his computer prowess (i.e., putting her outcomes in Ron's hands) or when she provides Ron with social support (i.e., taking Ron's outcomes into her own hands). *Attachment* increases when partners simply perceive greater inherent or intrinsic value in being together, quite apart from any objective considerations. This means that partners can express commitment through the symbolic value they attach to being together. For instance, Gayle can express her commitment to Ron when she comes to love him all the more despite her occasional annoyance with his extraversion and sloppiness (Murray & Holmes, 1993).

For mutually responsive interactions to develop, both partners must be equally willing to set aside self-interested concerns (Drigotas, Rusbult, & Verette, 1999). That is, partners must be equally committed. Any asymmetry in their commitment invites coordination difficulties (Drigotas et al., 1999; Sprecher, Schmeeckle & Felmless, 2006). Imagine the difficulty of coordinating mutually responsive interaction patterns in a marriage where Gayle is more committed to Ron than Ron is committed to her. Being less invested in the relationship gives Ron greater power and disproportionate license to behave selfishly and disappoint Gayle. Not needing Gayle as much as she needs him frees him from having to care about her reactions to his behavior. In such a marriage, Ron would face the constant temptation to be selfish, and Gayle would face the constant need to protect against his possible exploitation. Waller (1938) described this adaptive problem in terms of the *principle of least interest.* With unequal commitment, Waller reasoned, the power to be selfish resides with the partner who needs the relationship less (i.e., less invested), and the demand to sacrifice falls largely on the partner who needs the relationship more (i.e., more invested).

The if–then rules for motivating responsiveness minimize the chance of such power imbalances developing in the first place. Figure 8.1 illustrates these rules. The if–then rules coordinate partners' concrete and symbolic expressions of commitment by motivating different affective, cognitive, and behavioral reactions in low-risk and high-risk conflict-of-interest

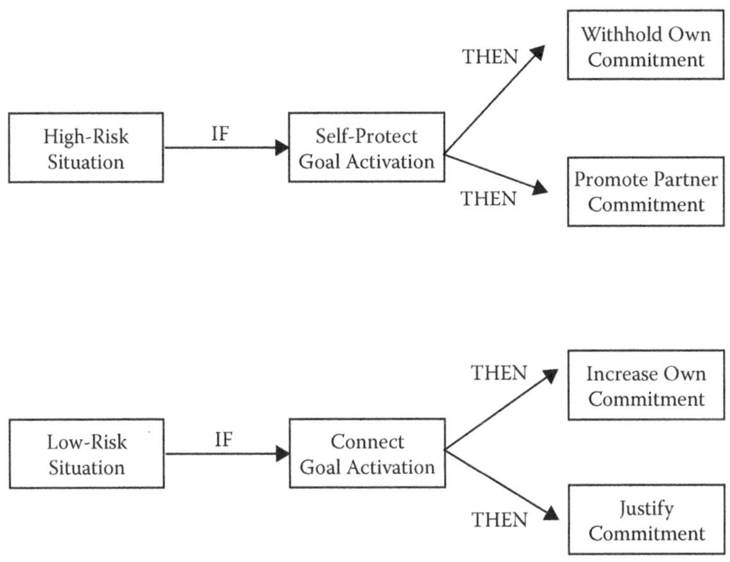

Figure 8.1 "If–then" procedural rules for motivating mutual responsiveness.

situations.[2] Consider a conflict-of-interest situation in which Gayle has golf plans on the Saturday Ron wants her to attend his family reunion. Imagine that Ron thinks that Gayle loves golfing for exercise and barely tolerates family reunions. Her perceived preferences make the situation high in the risk of her nonresponsiveness for Ron. In this situation, expecting Gayle to choose golf strengthens Ron's *state* goal to self-protect against the possibility of such a rejection. The activation of this goal (i.e., if) then elicits two complementary behavioral strategies (i.e., then) for averting such negative outcomes. Namely, it activates behavioral tendencies to withdraw from Gayle (i.e., withhold his own dependence) until he has taken some step that gives her little choice but to be responsive (i.e., promote her dependence). For instance, he might not ask her to attend (i.e., withhold his dependence) until he puts her in his debt by fixing a computer problem she could not solve on her own (i.e., promote her dependence).

Now imagine instead that Ron thinks that Gayle goes golfing only to make her friends happy and that she adores family reunions. Such preferences now make the situation low in the risk of her nonresponsiveness for Ron. In this situation, expecting Gayle's responsiveness strengthens his *state* goal to connect. The activation of this goal (i.e., if) then elicits two complementary behavioral strategies (i.e., then) for attaining such positive

[2] We have abstracted the elements of the Murray and Holmes (2009) model that are most central to our discussion here.

outcomes. Namely, it activates a behavioral strategy to escalate his dependence on her (e.g., asking her to skip golf and attend the reunion) and a compensatory strategy to justify any cost to his goal pursuits he incurs through his greater level of dependence and closeness (e.g., seeing Gayle's athleticism more positively when she wakes him to go running on the morning of the reunion).

The Interdependent Mind's Efficiency and Flexibility

The model of the interdependent mind proposed by Murray and Holmes (2009) also assumes that the if–then rules are efficient and flexible. By efficient, we mean that these rules are implicit procedural features of people's general working knowledge of relationships (Baldwin, 1992). Because the mind cannot afford the luxury of thinking through every decision it needs to make, social cognition scholars assume that ongoing and complex problems underlying social life have automatic and effortless solutions (Bargh, 2007; Bargh & Ferguson, 2000; Bargh & Williams, 2006; Dijksterhuis, Chartrand & Aarts, 2007; Dijksterhuis & Nordgren, 2006). Indeed, growing evidence suggests that the interdependent mind's if–then rules operate without conscious mediation. Situations that prime the if elicit the propensity to engage in the then without any conscious intent (Murray, Aloni, et al., 2009; Murray, Derrick, et al., 2008; Murray, Holmes, et al., 2009).

By *flexible*, we mean that people can correct or override the rules if they possess both the motivation and the opportunity to do so (Olson & Fazio, 2009). Models of attitudes, impression formation, and stereotyping assume that such automatic propensities control behavior unless people have the *motivation* and *opportunity* to override them (Fazio & Towles-Schwen, 1999; Fiske & Neuberg, 1990; Gilbert & Malone, 1995; Kunda & Spencer, 2003; Olson & Fazio; Wilson, Lindsey, & Schooler, 2000). Consistent with this logic, the behavioral effects of automatically activated goals can be overridden by situational cues that suggest pursuing such goals might preempt more important goal pursuits (Aarts, Custers, & Holland, 2007; Macrae & Johnston, 1998). For instance, people primed with the goal to be helpful automatically pick up *clean* pens for a clumsy colleague, but they decide to leave ink-stained pens at her feet because the conscious goal of staying clean trumps the unconscious goal to help (Macrae & Johnston). Applying this flexibility criterion to relationships means that partners can correct or overturn an automatic impulse to behave in a particular way if they are both motivated and able (Murray, Aloni, et al., 2009; Murray, Derrick, et al., 2008; Murray, Holmes, et al., 2009).

Why the Unconscious Might Better Sustain Commitment

The efficiency and flexibility of the interdependent mind suggest two reasons the unconscious mind might sometimes better sustain commitment than the conscious mind.

First, when Gayle is both motivated and able, she might correct for the automatic activation of if–then rules that might help sustain her commitment (Murray & Holmes, 2009). In this way, she might choose to overturn an adaptive behavioral impulse. For example, low in self-esteem, Gayle generally anticipates Ron's rejection—a painful possibility that she is highly motivated to avoid (Murray et al., 2006). Being *chronically* preoccupied with avoiding rejection therefore motivates Gayle to correct her *state* impulses to increase or justify her commitment on those occasions when she has the executive strength available to do so (Hofmann, Friese, & Strack, 2009). By keeping herself from feeling too committed, Gayle can essentially protect herself against the possibility of Ron's rejection in advance. In this way, Gayle's conscious doubts about Ron's responsiveness can interfere with her behaving on the basis of automatic impulses that might best sustain her commitment.

Second, the contents of Gayle's unconscious may be more diagnostic of the actual behavioral rewards that motivate her commitment than the contents of her conscious (Murray, Holmes, & Pinkus, 2010). Murray and Holmes (2011) argue that the interdependent mind keeps two separate barometers to gauge the partner's trustworthiness—one unconscious and one conscious. These barometers signal the safety of expressing commitment by foretelling the partner's anticipated responsiveness in newly emerging situations (Murray et al., 2006).[3]

The *unconscious* barometer of trust signals Ron's responsiveness through Gayle's overall automatic evaluative response to Ron—whether positive or negative. This ledger of the partner's responsiveness develops through processes of associative learning (Bargh & Huang, 2009; Bargh & Morsella, 2008; Murray, Holmes, & Pinkus, 2010). Gayle unconsciously learns whether connecting to Ron is safe in the process of interacting with him. This can happen in two ways. First, the simple exposure to relatively low- versus high-risk situations might condition Gayle's automatic evaluative association to Ron because Gayle learns that good versus bad things happen to her when she is in Ron's presence. That is, Ron's responsive and nonresponsive behavior conditions Gayle's more or less positive automatic evaluative association to Ron. Second, the if–then rules that control Gayle's reactions to Ron in relatively low- versus high-risk situations might also

[3] Holmes and Rempel (1989) define trust as the expectation that a partner can be counted on to meet one's needs, now and in the future. Expectations of responsiveness are thus central to the experience of trusting in another.

condition her automatic evaluative association to Ron because being in his presence makes her feel, think, or behave in particular ways. We elaborate on these points next by drawing on a longitudinal study of newlyweds (Murray, Holmes, & Pinkus, 2010).

In this study, we indexed situation-exposure and if–then habits of thought in the initial months of marriage through daily diary reports. Then we measured automatic evaluations of the partner (through the Implicit Associations Test) after 4 years of marriage. Situations and their associated if–then rule habits conditioned automatic evaluative associations to the partner. The more instances of Ron behaving nonresponsively (i.e., high-risk situations) Gayle encountered early on, the less positive her automatic evaluative response to Ron after 4 years. Similarly, the more often Gayle reacted to feeling rejected in such situations by distancing herself from Ron (i.e., a stronger "withhold own dependence" rule habit), the less positive her later automatic evaluative response. In contrast, the more often Gayle successfully compensated for costs by valuing Ron (i.e., a stronger "justify own commitment" rule habit), the more positive her later automatic evaluative associations to thinking of Ron.

The *conscious* barometer of trust signals Ron's responsiveness through Gayle's deliberative or considered reflections on Ron's likely motivations—whether positive or negative. While not immune to behavioral experience, this ledger of the partner's responsiveness develops largely through processes of appraisal and deliberative reasoning (Baldwin, 1992; Holmes & Rempel, 1989; Murray, Rose, Bellavia, Holmes, & Kusche, 2002; Murray et al., 2005; Wieselquist, Rusbult, Foster, & Agnew, 1999). With this metric, Gayle essentially decides whether connecting to Ron is safe by using her experiences to reflect on his likely feelings and beliefs about her (e.g., considering the qualities he values in her, the alternatives he might pursue). In her case, Gayle has come to see Ron's commitment as tenuous because she incorrectly assumes that he sees the same faults in her that she sees in herself (Murray et al., 2000). Similarly, her projection of her uneasy relationship with her parents onto her marriage to Ron (Mikulincer & Shaver, 2003) further exacerbates her doubts about his commitment to her.

Therefore, for a low-self-esteem Gayle, relatively unconscious, positive associations to Ron might sustain her commitment better than her uncertain conscious expectations because unconscious sentiments better capture the actual rewards that sustained her commitment in the past. The experiential basis of Gayle's automatic positive associations to Ron might then give her unconscious the power to sustain her commitment in the face of Ron's negative or rejecting behaviors. In a sense, her automatic evaluative association to Ron might function as a "selfish" goal to connect, by monopolizing her perceptions and behaviors (Bargh & Huang, 2009). Bargh and Huang define a "selfish" goal as one that pursues its own agenda, independent of other situational contingencies (Bargh, Gollwitzer & Oettingen, 2010). Once active, the selfish goal "takes the helm of one's mental machinery and adjusts its settings in a single-minded way to maximize the chances of goal attainment" (Bargh et al., p. 300). In Gayle's case, positive evaluative associations to Ron could create a chronic pull to be close to Ron (i.e., a selfish goal) that makes her less likely to distance from him in reaction to her conscious reservations about his responsiveness (Murray, Pinkus, et al., 2011). As a result, possessing a positive automatic evaluative response toward Ron could help sustain her commitment in the face of his transgressions (e.g., breaking a promise, his criticisms of her).

For both of the reasons previously outlined, the unconscious mind sometimes might better sustain Gayle's commitment than her conscious mind. We elaborate on each of the points already raised as we now turn to examine how the conscious and unconscious minds jointly manage the threats to commitment posed by autonomy costs and the partner's rejecting behaviors.

How the Unconscious Compensates for Autonomy Costs

In becoming more committed and objectively more dependent on the partner, people inevitably incur costs. When she draws closer to Ron, Gayle loses something important. She loses the autonomy to pursue her own goals without interference from Ron. For instance, she loses some of her personal autonomy to pursue alternate friends or even quiet solitude because Ron wants her to spend as much time with him and his friends as possible. She also has to suffer through his fallibility and unintentional interference with her goals. Being connected to him means that she has to tolerate his seasonal obsessions with basketball, his disparate musical tastes, and his snoring. Because drawing closer to Ron can have such unintended costs for Gayle's autonomous pursuit of her goals, the interdependent mind automatically deflects attention from such costs. Without such a mechanism, the costs that come from becoming closer and more committed to Ron could eventually undermine Gayle's commitment and motivation to be responsive to his needs (Clark & Grote, 1998).

The "justify own commitment" rule introduced in Figure 8.1 keeps commitment on track in the face of autonomy costs. This if–then rule maintains Gayle's resolve in her commitment by making Ron more valuable precisely when he is more costly (Murray et al., 2009). By turning adversity into statements of the partner's inherent value (Higgins, 2006), such an impulse sustains one's motivation to be responsive by protecting the newly exercised state of commitment. The inherently reciprocal nature of relationships makes this rule a functional one: Meeting a partner's needs elicits that partner's willingness to meet one's own needs (Clark &

Grote, 1998; Holmes, 2002; Kelley, 1979; Reis et al., 2004). Maintaining one's commitment to meeting the partner's needs thereby functions as a safety check to ensure the partner's general and reciprocated motivation to be responsive in future situations.

As we would expect, making the partner's fallibility salient automatically elicits behavioral intentions to protect the relationship bond. For instance, imagining that a close other forgot to mail one's application for a coveted job opportunity immediately brings thoughts of forgiving the errant partner to mind (Karremans & Aarts, 2007). This reflexive tendency to compensate for a partner's fallibility is so powerful that people respond with positive facial affect to signs of a significant other's faults in a novel other (Andersen, Reznik, & Manzella, 1996). Making salient the opportunity costs that come from foregoing attractive alternative partners also automatically elicits commitment-bolstering behaviors. Lydon and his colleagues argue that women in committed relationships possess a built-in if–then implementation intention for staving off temptation (Lydon, Menzies-Toman, Burton, & Bell, 2008). In one experiment, they primed feelings of connection to one's romantic partner and then measured how close participants allowed themselves to be to a tempting alternative partner in a virtual reality environment. When primed with connection, women, but not men, actually moved further away from virtual temptation. Women's implementation intention to shut their eyes to alternatives proved to be so engrained that simply chatting with an attractive alternative partner increased their willingness to tolerate their own partner's transgressions. Even men evidenced such defenses when the circumstances were right. Men trained in the implemental intention to distract themselves from costs turned away from a tempting virtual alternative partner.

When the Conscious Mind Gets in the Way

Despite the commitment-sustaining benefits of compensating for the partner's fallibility (Murray, 1999), the unconscious and conscious mind can disagree in how to handle autonomy costs. The Motivation and Opportunity as Determinants (MODE) model of the attitude-behavior relation (Fazio & Towles-Schwen, 1999) assumes that people can overturn the automatic impulse to behave in a given way if they both want to overturn the impulse (i.e., motivation) and if they have the necessary cognitive resources (i.e., opportunity) available to do so (Olson & Fazio, 2009). Trust in the partner's commitment supplies the motivation to correct *state* goals that conflict with broader goal pursuits. For Gayle, being less trusting heightens the potential hurts that could come from valuing Ron more (Murray, Griffin, Rose, & Bellavia, 2003). For this reason, Gayle's broader and *chronic* goal to self-protect can motivate her to curb the state goal to connect to Ron when it arises. However, correction also requires opportunity. This means that even if Gayle is motivated to self-protect, she can correct her automatic impulse to value Ron only when she has the cognitive resources (i.e., executive strength) available to stop herself from valuing him (Gilbert & Malone, 1995; Muraven & Baumeister, 2000; Murray et al., 2008). Indeed, people who are higher in the capacity to allocate and control attention have an easier time overturning their relatively unconscious impulses. For instance, people high in working memory capacity overturn automatic impulses to eat the chocolate M&M's they crave and eat only as few as their intention to diet dictates (Hofmann, Gschwendner, Friese, Wiers, & Schmitt, 2008).

The logic of the MODE model suggests that the conscious mind should interfere with the unconscious mind's efforts to sustain commitment when people are both motivated and able to correct their in-the-moment automatic impulses. We directly examined this possibility in a series of studies that investigated how people sustain or ensure commitments in the face of autonomy costs (Murray et al., 2009). In designing these studies, we drew on basic principles of social cognition research. Such research distinguishes automatic and controlled processes by employing manipulations and measures that differ in the level of deliberation they afford (Bargh, 1994). Consider the distinction between implicit and explicit measures of prejudice. The Implicit Associations Test (IAT) examines how quickly people make "congruent" (i.e., White = good, African American = bad) versus "incongruent" (i.e., White = bad, African American = good) responses (Greenwald, McGhee, & Schwartz, 1998). The IAT is thought to capture an automatic form of prejudice because people cannot decide to speed up their reactions to incongruent trials to mask their prejudices. In contrast, the Feeling Thermometer asks people to rate how positively they feel toward different racial groups (Haddock, Zanna, & Esses, 1993). It captures a relatively controlled form of racial prejudice because people can deliberately decide to evaluate a minority group member positively rather than negatively.

We use similar logic to distinguish automatic and controlled processes in our research on commitment insurance. Namely, we assumed that manipulations and measures that limit the opportunity for deliberation should reveal the influence of the unconscious. On such measures, the reactions of a low- and high-self-esteem person (who are less and more trusting, respectively) should be hard to distinguish. Both should be quicker to "justify commitment" when autonomy costs are salient. But their responses should diverge on manipulations and measures that afford the conscious sufficient opportunity (i.e., executive control) to correct. On such measures, a low-self-esteem Gayle might actually withdraw her commitment because she has the chronic motivation (and now the acute opportunity) to correct the impulses of her unconscious mind.

We tested these hypotheses through a combination of laboratory experiments and field studies (Murray et al., 2009). In two experiments, we used laboratory manipulations to make the ways their partner interfered with their goals salient to our dating participants (Murray et al.). We then measured how the unconscious and the conscious mind manage such autonomy costs. Specifically, we measured the unconscious mind's defensive response by examining how quickly participants associated their partner with positive traits—an implicit measure of commitment justification over which participants have limited executive control. We measured the conscious mind's defensive response by examining participants' self-reported positive illusions—an explicit measure of commitment justification over which participants have substantial executive control. In these studies, we captured the motivation to correct the if–then rules through global self-esteem. We used global self-esteem as a proxy for trust because low-self-esteem people typically expect their partner to be less responsive to their needs than high-self-esteem people (Murray et al., 2000).

In our first experiment, we primed the costs of connection indirectly. We asked participants in the autonomy costs priming condition to think back to a time when they complained about their relationship to a friend. Once they had this conversation in mind, we then asked participants to describe the two biggest costs their friend pointed out in their dating relationship (e.g., partner is too jealous, too controlling, too lazy). Participants in the normative costs control condition described two costs their friend perceived in most relationships. Baseline control participants did not describe any costs. We then measured the automatic (i.e., unconscious) and controlled (i.e., conscious) tendency to justify commitment by evaluating the partner all the more positively. We tapped automatic compensation by examining how quickly people associated their dating partner with positive traits in a person/object categorization task. We tapped controlled compensation by examining how people decided to describe their relationship on explicit measures assessing perceptions of the partner's traits, perceptions of control over relationship conflicts, and optimism about the future (Murray & Holmes, 1997).

In a second experiment, we primed autonomy costs directly. We had participants in the autonomy costs priming condition complete an electronic survey (which we had biased to suit our purposes). This purportedly diagnostic instrument took participants through a grocery list of the common ways the partner could thwart one's personal goals. The caption at the top of this "diagnostic instrument" instructed participants to indicate whether any of the events had happened to them because of their relationship with their partner (e.g., "I couldn't watch something I wanted to watch on TV"; "I had to go out with friends of my partner that I didn't like"; "I had my sleep disrupted"). Participants uniformly indicated that many if not most of these negative events had happened to them. Then they received "diagnostic" feedback that their involvement in the relationship had caused them to make so many changes that they no longer had much personal control over their life. Participants in the thwarted outcome control condition completed a survey that simply asked them to indicate whether specific events where they could not meet their goals had ever happened to them (e.g., "I couldn't watch something I wanted to watch on TV"; "I had my sleep disrupted"). Participants in the baseline control condition did not complete a survey. We then administered the measures of automatic and controlled partner valuing (our indices of commitment justification in these experiments).

Figure 8.2 presents the results for the implicit measure of partner valuing for each experiment. As we expected, participants primed with autonomy costs justified their commitments in both experiments. They were actually quicker to associate their partner with positive traits than control participants. That is, being primed with costs automatically caused them to value their partner more. The automatic tendency to compensate for costs by reflexively associating the partner with positive traits emerged regardless of self-esteem (our dispositional proxy for trust). Figure 8.3 presents the results for the explicit measure of commitment

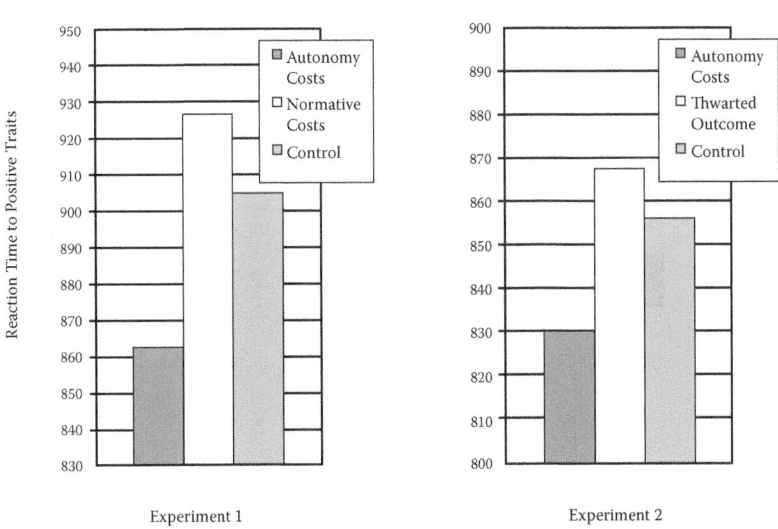

Figure 8.2 Implicit evaluations of the partner as a function of autonomy costs (quicker reaction times to positive traits indicate more positive implicit evaluations). (From Murray, S.E., & Holmes, J.G., *Interdependent Minds: The Dynamics of Close Relationships*, Guilford Press, New York, 2011. Copyright Guilford Press. Reprinted with permission from Guilford Press.)

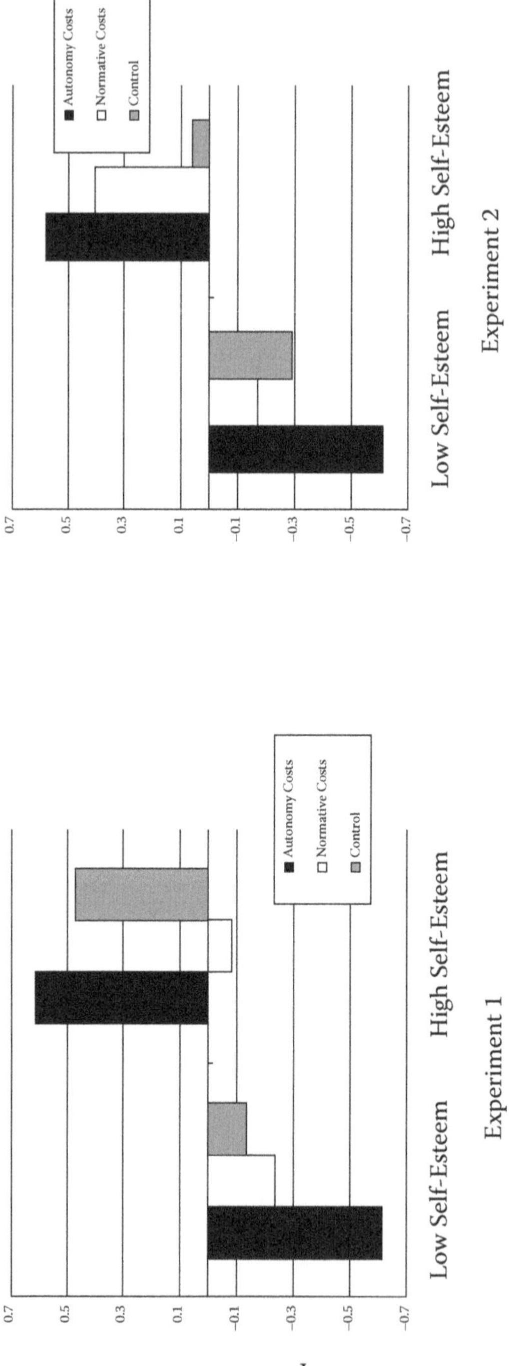

Figure 8.3 Explicit positive illusions as a function of autonomy costs and self-esteem. (From Murray, S. L., Pinkus, R.T., Holmes, J.G., Harris, B., Gomillion, S., Aloni, M., et al., *Journal of Personality and Social Psychology*, 101, 485–502, 2011. Reprinted with permission from the American Psychological Association.)

justification in both experiments. By drawing people's conscious attention to their partner, these self-report scales afforded less trusting people the opportunity to deliberate and correct their automatic responses. Low-self-esteem people jumped on this chance. In both experiments, cost-primed lows expressed less positive evaluations of their partner, less efficacy, and less optimism on the explicit measure of positive illusions. Their conscious sentiments thus disputed their unconscious sentiments. No such contrary conscious mind emerged for high-self-esteem people. Instead, high-self-esteem people also justified their commitment on the explicit measures.

A daily diary study of newlywed marriages also revealed that the capacity to act on one's automatic motives and compensate for costs has important real-world consequences. In this study, newlyweds completed an electronic diary each day for 14 consecutive days. Each day they indicated how often their partner's actions interfered with their personal goals that day (e.g., Ron doing something he wanted to do rather than what Gayle wanted to do; Ron using the last of something Gayle needed; Ron refusing to talk about something Gayle wanted to discuss). They also reported on how much they valued their partner each day. When we analyzed the data without taking the motivation to correct into consideration, we found a main effect of autonomy costs. These newlyweds generally compensated for one day's costs by valuing their partner more on the next day. But this picture changed dramatically once we took the motivation to correct into consideration (again captured by using self-esteem as a proxy for trust). Figure 8.4 presents the results. High-self-esteem people valued their partner more on days after their partner thwarted more of their goals. Low-self-esteem people did not; they instead curbed this relationship-protective impulse. They did so to the detriment of their partner and relationship.

Valuing a costly partner more should sustain one's commitment and, in so doing, sustain one's motivation to be responsive to the partner's needs. In the diary study, Ron indicated how responsively Gayle behaved toward him (e.g., "Gayle listened to and comforted me"; "Gayle helped me solve a problem I was having"; "Gayle did something she didn't really want to do because I wanted to do it"; "Gayle put my tastes ahead of her own"). Analyses revealed that when Ron had imposed costs on Monday and Gayle compensated for these costs by valuing him more on Tuesday, Ron actually perceived Gayle as behaving more responsively on Tuesday as well (Murray et al., 2009). By failing to compensate for the costs their partner imposed, low-self-esteem people effectively undermined the quality of their interactions: They overturned an automatic inclination that could have sustained their willingness to be responsive to their partner's needs.

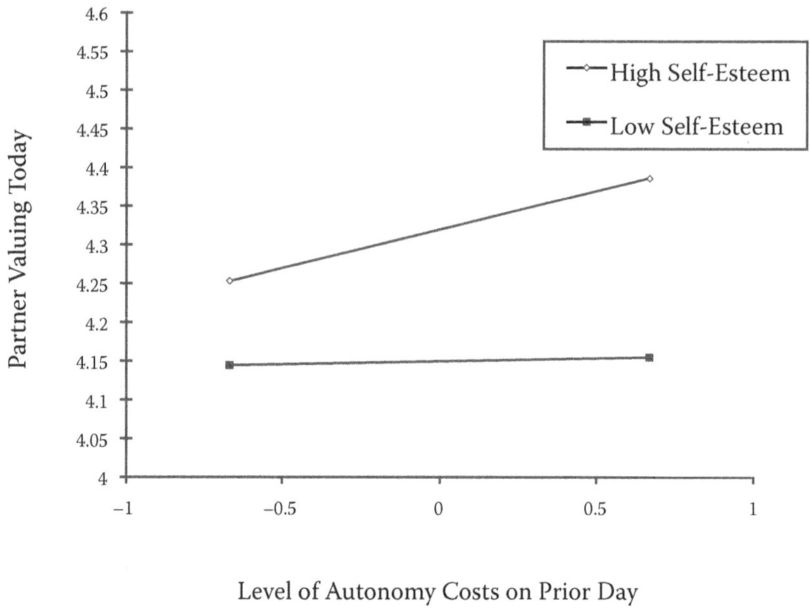

Figure 8.4 Compensating for daily autonomy costs as a function of self-esteem. (From Murray, S. L., Pinkus, R. T., Holmes, J. G., Harris, B., Gomillion, S., Aloni, M., et al., *Journal of Personality and Social Psychology*, 101, 485–502, 2011. Reprinted with permission from the American Psychological Association.)

A 1-year follow-up on our newlyweds revealed that the long-term consequence of ignoring the unconscious was even more serious. The daily diary data allowed us to derive an index of cost compensation for each person. This within-person slope residual indexes how much (or how little) each participant values the partner more in response to costs. We then used this cost-compensation index to track the long-term effects of failing to compensate for costs. Namely, we predicted changes in satisfaction over the year of marriage from initial levels of self-esteem, from cost compensation, and from both self-esteem and cost compensation. These successive analyses revealed evidence of mediation. Namely, newlyweds with lower self-esteem reported greater declines in satisfaction over the year. Newlyweds who compensated less for costs also reported relatively greater declines in satisfaction. Further pointing to a pattern of mediation, once we controlled for the tendency to compensate for costs, the association between self-esteem and later satisfaction was substantially reduced. In sum, low-self-esteem people reported less satisfaction over the first year of marriage partly because their conscious mind would not let them compensate for costs their partner imposed.

How the Unconscious Counters Rejection Anxiety

In becoming more committed and objectively more dependent on the partner, people inevitably invite disappointment as well. No matter how good his intentions in general, Ron will undoubtedly behave in ways that give Gayle reason to worry about his rejection (Murray & Holmes, 2009). He might make an off-handed criticism about her weight or forget to do the dishes as promised. Low in self-esteem, Gayle is also likely to perceive rejection in events as innocuous as Ron simply being in a bad mood (Bellavia & Murray, 2003). Typically, anticipating a partner's rejection motivates people to distance themselves from the source of the hurt. In fact, extensive research on risk regulation processes suggests that anticipating a partner's rejection generally limits expressions of commitment (Murray et al., 2006). When people question their partner's responsiveness, they cannot afford to feel truly close and committed. Instead, they minimize the likelihood and pain of rejection by withdrawing (Murray et al., 2006), both practically (e.g., seeking less support) and symbolically (e.g., valuing the partner less). Consequently, such conscious or deliberated concerns about the partner's responsiveness erode satisfaction and commitment in dating and marital relationships (Downey, Freitas, Michaelis, Khouri, 1998; Murray, Bellavia, Rose, & Griffin, 2003; Murray, Griffin, et al., 2003; Murray et al., 2000; Murray, Holmes, Griffin, Bellavia, & Rose, 2001; Wieselquist et al., 1999).

Might the unconscious mind know something about the actual rewards of interacting that could keep conscious rejection anxieties from eroding commitment? There are at least four reasons to attribute such "selfish" power to automatic evaluative associations to the partner.

First, as we reasoned earlier, automatic evaluative associations might better capture the actual structure of rewards (and punishments) that have motivated connection-seeking (or self-protective) behaviors in the past (Kelley, 1979). Second, more positive automatic evaluations of the partner generally might make the partner's behavior seem less rejecting. Such associations might function as a positive filter that biases the perception of incoming information (Banaji & Heiphetz, 2010; Bargh et al., 2010; Dijksterhuis, 2010; Fazio, 1986). Because interacting with the partner automatically primes one's evaluative response toward the partner, negative behaviors in specific situations are likely to be assimilated to more positive general evaluations (Kunda & Spencer, 2003; Mikulincer, Hirschberger, Nachmias, & Gillath, 2001). Therefore, positive automatic evaluative associations might make a partner's transgressions less hurtful and, as a result, less likely to motivate one's own avoidance of the partner.

Third, automatic evaluative associations are likely to be slow to align or catch up with newly emerging doubts that could compromise such positive connection-promoting sentiments. Automatic attitudes are generally thought to be asymmetrically malleable, more readily formed than undone (Gregg, Seibt, & Banaji, 2006). In contrast, deliberative expectations about a partner's regard, love, and commitment are highly sensitive to circumstance (Murray, Holmes, MacDonald, & Ellsworth, 1998; Murray et al., 2002; Murray, Bellavia et al., 2003). Such logic implies that positive automatic evaluative associations might provide a counterweight to emerging doubts because such sentiment stubbornly retains its positivity. Fourth, automatic evaluative associations can elicit corresponding behaviors even when contradictory explicit sentiments are accessible in memory (Wilson et al., 2000). For instance, unconsciously primed thoughts of security heighten empathy for others (Mikulincer, Gillath, Halevy, Avihou, Avidan, & Eshkoli, 2001), diminish out-group derogation (Mikulincer & Shaver, 2001), and increase the desire to seek support from others in dealing with a personal crisis (Pierce & Lydon, 1998) even when people's considered expectations about others oppose such behaviors. This logic implies that positive automatic evaluative associations might motivate sustained expressions of commitment even when people have reason to be cautious.

For these reasons, we expected people who evidence more positive automatic evaluative associations to their partner to be better equipped to sustain their commitment in the face of rejection anxieties. That is, we expected them to continue to value their partner even when they had reason *not* to do so. Nonetheless, the conscious mind might get in the way of the unconscious mind's efforts to sustain commitment when people have greater cognitive capacity available to listen to and heed their conscious doubts. Laboratory manipulations (e.g., rehearsing an alphanumeric string, suppressing emotion) are often used to deplete cognitive capacity (Muraven & Baumeister, 2000). However, people also differ chronically in their capacity to self-regulate because they differ in working memory capacity (Hofmann et al., 2008). Namely, people who are low in working memory capacity possess limited cognitive capacity to shield information in working memory from distraction or interference (Baddeley & Hitch, 1974; Hofmann et al.). Being low in working memory capacity thus disposes people to act on the basis of automatic impulses because they often lack the cognitive resources needed to override their impulses. On this basis, we expected more positive automatic evaluative associations to the partner to better sustain commitment for people who were low in working memory capacity because they have less chronic opportunity to correct their automatic impulses.

We tested these hypotheses through a combination of laboratory experiments and field studies (Murray, Pinkus, et al., 2011). In one of our experiments, we brought dating couples into the laboratory. First, participants completed the IAT to measure automatic evaluative associations to their partner (Greenwald et al., 1998; Zayas & Shoda, 2005). In our version of the IAT, participants categorized words belonging to four categories: (1) pleasant words (e.g., vacation, pleasure), (2) unpleasant words (e.g., bomb, poison), (3) words associated with the partner, and (4) words not associated with the partner (Zayas & Shoda). We contrasted reaction times on two sets of trials to diagnose automatic associations to the partner. In one set of trials, participants used the same response key to respond to pleasant words and partner words (i.e., compatible pairings). In the other set of trials, participants used the same response key to respond to unpleasant words and partner words (i.e., incompatible pairings). The logic of the IAT says that people who evidence more positive automatic associations to the partner should be faster when categorizing words using the same motion for "partner" and "pleasant" than when using the same motion for "partner" and "unpleasant."

We then gave the conscious mind reason to fear the partner's rejection for participants in the experimental but not the control condition. We led target participants in the *rejection* condition to believe that their partner was spending a copious amount of time providing a long list of the participant's faults (when the partner was actually listing 25 items in their residence). Participants in the *control* condition were led to believe their partner listed only one or two faults. Normally, "seeing" one's partner itemize numerous complaints leaves people feeling hurt and rejected (Murray et al., 2002). However, we expected that being armed with a more positive automatic association to the partner would take the sting away from this hurtful experience in a way that helped sustain commitment. To look at the strength of people's commitment, we measured how quickly experimental and control participants associated their partner with positive traits in a subsequent categorization task—a measure of partner valuing that captures psychological attachment. The results confirmed our hypotheses. For people who evidenced *less* positive automatic evaluative associations to their partner on the IAT, thinking their partner had a laundry list of complaints automatically motivated partner devaluing. That is, they were significantly slower to associate their partner with positive traits than control participants. For people who evidenced *more* positive automatic evaluative associations on the IAT, thinking their partner had a laundry list of complaints had absolutely no effect. That is, these "protected" participants were just as quick to associate their partner with positive traits as control participants. Thus, having a more positive automatic evaluative association to Ron kept Gayle's commitment intact even when he seemed to be compiling a mammoth list of her faults.

When the Conscious Mind Gets in the Way

The logic of correction we described earlier suggests that the unconscious mind should lose its power to sustain commitment when people possess more of the executive control needed to heed their conscious doubts. In a second experiment, we examined this possibility (Murray, Pinkus, et al., unpublished data). In this study, we measured automatic associations to the partner using the IAT, and we measured working memory capacity using a computation span task (Hofmann et al., 2008). In this task, participants try to remember the results of algebraic equations (e.g., 2 + 3 = 7, 6 − 4 = 2) while simultaneously deciding whether each equation is true or false. People who are low in working memory capacity perform more poorly on this dual demand task. Next we led participants in the *rejection-salient* condition to believe that their partner had a yet to be proclaimed complaint about some aspect of the participant's personality or behavior. Then we measured partner valuing, a measure of commitment strength, by examining how quickly experimental and control participants associated their partner with positive traits in a reaction-time task.

Figure 8.5 presents the results. As we expected, the protective effects of having more positive automatic evaluative associations to the partner were evident only for people low in working memory capacity (the left-hand side of the figure). People who were low in working memory capacity and evidence more positive automatic evaluative associations sustained commitment in the face of the partner's anticipated rejection.

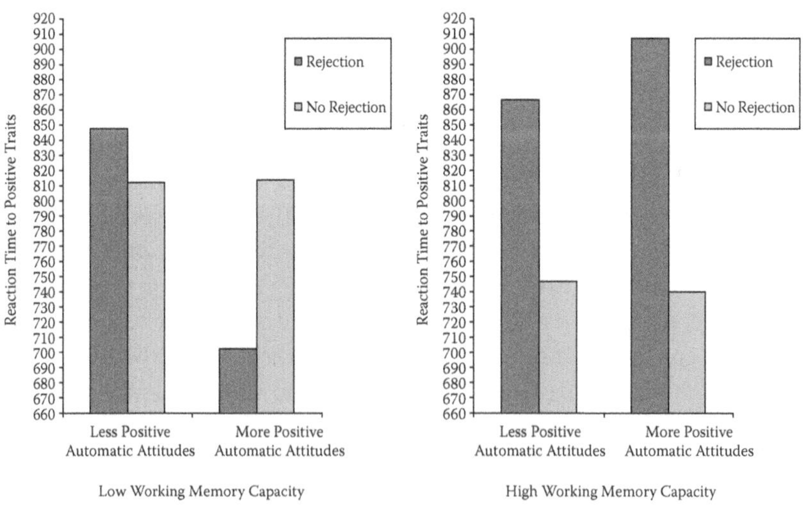

Figure 8.5 Implicit partner valuing as a function of rejection salience, automatic partner attitudes, and working memory capacity.

That is, they were just as quick to associate their partner with positive traits as control participants. However, people who were low in working memory capacity and evidenced less positive automatic associations to the partner distanced themselves from a seemingly rejecting partner. That is, they were slower to associate their partner with positive traits than control participants. In contrast, people who were high in working memory capacity seemed to ignore the commitment-sustaining inclinations of their unconscious. They were slower to associate their partner with positive traits when they consciously ruminated about their partner's rejection. This was true regardless of their automatic evaluative associations to the partner as assessed by the IAT.

We found further evidence of such dynamics in a correlational study of married couples (Murray & Holmes, 2011). In this study, we again measured automatic evaluative associations to the partner and working memory capacity. Rather than experimentally manipulating rejection anxiety, we measured chronic rejection anxiety through measures of trust, perceived commitment, and perceptions of the partner's regard for one's traits. New to this study, we also used a behavioral measure of sustained commitment—resisting the impulse to be rejecting and nonresponsive. For example, we asked Ron how often Gayle engaged in cold, hostile, and rejecting behaviors toward him in the prior 2 weeks (e.g., "My partner snapped or yelled at me"; "My partner criticized/insulted me"; "My partner ignored/did not pay attention to me").

We again expected the commitment-sustaining benefits of having more positive automatic evaluative associations to be most evident for people low in working memory capacity. Figure 8.6 presents the results. This figure presents the association between automatic (i.e., unconscious) evaluations of the partner and behavioral distancing as a function of two factors: (1) conscious expectations of partner responsiveness; and (2) working memory capacity. The results for people who are low in working memory capacity (the left side of Figure 8.6) are the most telling.

People who were low in working memory capacity and evidenced a less positive automatic evaluative association toward their partner distanced themselves from a seemingly rejecting partner: That is, the greater their anxiety about their partner's rejection, the more distant, cold, and rejecting their behavior (as reported by their partner). However, people who were low in working memory capacity and evidenced a more positive automatic evaluative association toward their partner sustained commitment by not engaging in such rejecting behavior. In this case, Gayle's conscious doubts about Ron's rejection lost all of their power to compel her to distance and treat him badly (fortunately for Ron). None of the effects of automatic evaluative associations for people high in working memory capacity were

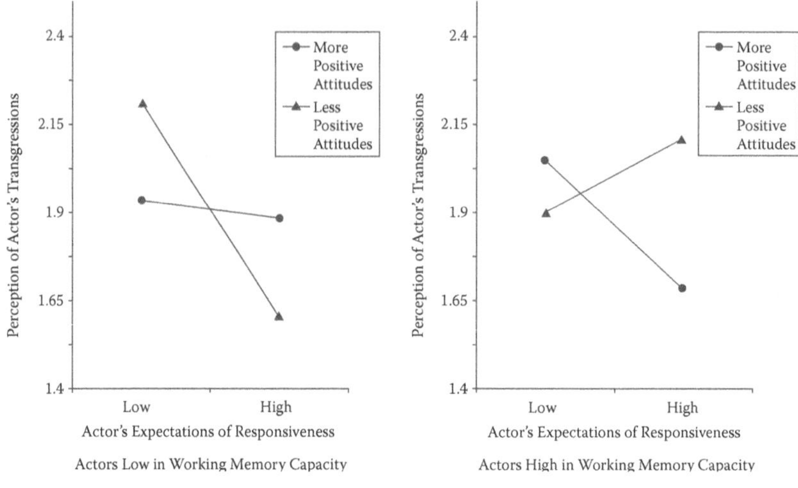

Figure 8.6 Partners' perception of actors' transgressions as a function of actors' automatic attitudes, working memory capacity, and expectations of responsiveness. (From Murray, S. E., & Holmes, J. G., *Interdependent Minds: The Dynamics of Close Relationships*, Guilford Press, New York, 2011. Copyright Guilford Press. Reprinted with permission from Guilford Press.)

significant. As in the experimental studies, then, having more positive automatic evaluative associations to the partner that one has little capacity to correct helps to sustain commitment in the face of conscious doubts about the partner's rejection.

A Dual-Motives Perspective on Relationships

The unconscious mind sometimes seems better able to protect one's commitment than the conscious mind. When left to the devices of the unconscious mind, people automatically come to value a costly partner more—a commitment-bolstering response that can increase the motivation to be responsive to the partner's needs. People who evidence more positive automatic evaluative associations to their partner also resist the impulse to distance themselves behaviorally from a seemingly rejecting partner—a commitment-bolstering response that can keep negative reciprocity cycles from developing (Murray, Bellavia et al., 2003).

The research reviewed in this chapter thus suggests that the unconscious mind might sometimes be better equipped to sustain commitment than the conscious mind. In making this point, we are not arguing that the conscious mind has no role to play in sustaining commitment. Considerable research suggests that people consciously defend and maintain their commitments (see Rusbult & Van Lange, 2003 for a review). In fact, people who face greater pressure to consciously justify their commitments because they are highly invested and have few alternatives to staying in the relationship claim to be happy and satisfied even when their automatic reactions to their partner are relatively negative (Scinta & Gable, 2007).

The point we wish to make is *not* that conscious defenses are ineffective (or that the conscious mind is nothing more than a relationship curmudgeon). Our point is that consciousness does not tell the whole story. For instance, more positive automatic evaluations of a dating partner predict greater relationship stability even when conscious feelings of satisfaction are taken into account (LeBel & Campbell, 2009; Lee, Rogge, & Reis, 2010). Such findings suggest that there is insight to be gained in considering how the unconscious and conscious jointly regulate commitment because unconscious and conscious sentiments can inform and contradict one another. Think of our low-self-esteem Gayle. When Ron interferes with her goals, her automatic inclination is to value him more. In situations where she is tired or distracted, she acts on this inclination and finds herself laughing as she trips over his mislaid shoes on the kitchen floor. However, in situations where she has greater self-regulatory capacity available, being consciously less trusting motivates her to correct her commitment-preserving impulses. In such situations, she instead criticizes Ron for the same mislaid shoes. In this way, the dissociation between Gayle's unconscious and conscious sentiments creates instability in her behavioral expressions of commitment. In fact, the sometimes warm, sometimes cold nature of Gayle's behavior might also induce instability in Ron's commitment (because he is never sure when to trust her to be responsive to his needs).

Now imagine the further complexity that Gayle's automatic evaluative association to Ron is highly positive (because she encountered multiple low-risk situations with him in the past). Contrast Gayle with Barbara, who is not only distrusting but also has a relatively negative automatic evaluative association to her partner Barney. Now imagine the various ways Ron (or Barney) might transgress over the course of a month (e.g., being critical, breaking a promise, not being supportive). Because her overall *automatic* evaluative response to Ron is positive, Gayle is better equipped to resist self-protectively distancing herself from Ron than Barbara is to resist such behaviors. Consequently, Gayle's interactions with Ron might be less likely to degrade into patterns of reciprocated negative affect and behavior (because her positive automatic associations to Ron help sustain her motivation to be close and committed to him).

Conclusion

The research reviewed in this chapter suggests that a smart relationship unconscious sometimes better knows how to sustain commitment.

The possibility of such a knowing unconscious suggests that vulnerable people, such as those low in self-esteem or insecure in attachment style, are not necessarily fated to relationship distress. Instead, when left to the devices of their unconscious, their relationship behavior may be just as adaptive as the behavior of more resilient people (Murray, Pinkus, et al., 2010). The possibility of a knowing unconscious guiding behavioral interaction further suggests that what partners cannot directly tell themselves (or close relationships researchers) may be even more important in diagnosing a relationship's fate than what partners can directly tell themselves (or close relationships researchers).

References

Aarts, H., Custers, R., & Holland, R. W. (2007). The non-conscious cessation of goal pursuit: When goals and negative affect are coactivated. *Journal of Personality and Social Psychology, 92*, 165–178.

Andersen, S. M., Reznik, I., & Manzella, L. M. (1996). Eliciting facial affect, motivation, and expectancies in transference: Significant-other representations in social relations. *Journal of Personality and Social Psychology, 71*, 1108–1129.

Baddeley, A. D., & Hitch, G. J. (1974). Working memory. In G. A. Bower (Ed.), *The psychology of learning and motivation* (Vol. 8, pp. 47–90). New York: Academic Press.

Baldwin, M. W. (1992). Relational schemas and the processing of social information. *Psychological Bulletin, 112*, 461–484.

Banaji, M. R., & Heiphetz, L. (2010). Attitudes. In S. T. Fiske, D. T. Gilbert, & G. Lindzey (Eds.), *Handbook of social psychology* (5th ed., Vol. 1, pp. 353–393). Hoboken, NJ: John Wiley & Sons, Inc.

Bargh, J. A. (1994). The four horsemen of automaticity: Awareness, intention, efficiency, and control in social cognition. In R. S. Wyer & T. K. Srull (Eds.), *Handbook of social cognition* (Vol. 1, pp. 1–40). Hillsdale, NJ: Erlbaum.

Bargh, J. A. (2007). *Social psychology and the unconscious: The automaticity of higher mental processes*. New York: Psychology Press.

Bargh, J. A., & Ferguson, M. J. (2000). Beyond behaviorism: On the automaticity of higher mental processes. *Psychological Bulletin, 126*, 925–945.

Bargh, J. A., Gollwitzer, P. M., & Oettingen, G. (2010). Motivation. In S. T. Fiske, D. T. Gilbert, & G. Lindzey (Eds.), *Handbook of social psychology* (5th ed., Vol. 1, pp. 268–316). Hoboken, NJ: John Wiley & Sons, Inc.

Bargh, J. A., & Huang, J. Y. (2009). The selfish goal. In G. B. Moskowitz & H. Grant (Eds.), *The psychology of goals* (pp. 127–150). New York: Guilford Press.

Bargh, J. A., & Morsella, E. (2008). The unconscious mind. *Perspectives on Psychological Science, 3*, 73–79.

Bargh, J. A., & Williams, E. L. (2006). The automaticity of social life. *Current Directions in Psychological Science, 15*, 1–4.

Baumeister, R. F., & Leary, M. R. (1995). The need to belong: Desire for interpersonal attachments as a fundamental human motivation. *Psychological Bulletin, 117*, 497–529.

Bellavia, G., & Murray, S. L. (2003). Did I do that? Self-esteem-related differences in reactions to romantic partners' moods. *Personal Relationships, 10*, 77–95.

Clark, M. S., & Grote, N. K. (1998). Why aren't indices of relationship costs always negatively related to indices of relationship quality? *Personality and Social Psychology Review, 2*, 2–17.

Dijksterhuis, A. (2010). Automaticity and the unconscious. In S. T. Fiske, D. T. Gilbert, & G. Lindzey (Eds.), *Handbook of social psychology* (5th ed., Vol. 1, pp. 228–267). Hoboken, NJ: John Wiley & Sons, Inc.

Dijksterhuis, A., Chartrand, T. L., & Aarts, H. (2007). Effects of priming and perception on social behavior and goal pursuit. In J. A. Bargh (Ed.), *Social psychology and the unconscious: The automaticity of higher mental processes* (pp. 51–132). New York: Psychology Press.

Dijksterhuis, A., & Nordgren, L. F. (2006). A theory of unconscious thought. *Perspectives on Psychological Science, 1*, 95–109.

Downey, G., Freitas, A. L., Michaelis, B., & Khouri, H. (1998). The self-fulfilling prophecy in close relationships: Rejection sensitivity and rejection by romantic partners. *Journal of Personality and Social Psychology, 75*, 545–560.

Drigotas, S. M., Rusbult, C. E., & Verette, J. (1999). Level of commitment, mutuality of commitment, and couple well-being. *Personal Relationships, 6*, 389–409.

Enfield, N. J., & Levinson, S. C. (2006). *Roots of human sociality: Culture, cognition and interaction*. New York: Berg.

Fazio, R. H. (1986). How do attitudes guide behavior? In R. M. Sorrentino & E. T. Higgins (Eds.), *The handbook of motivation and cognition: Foundations of social behavior* (pp. 204–243). New York: Guilford Press.

Fazio, R. H., & Towles-Schwen, T. (1999). The MODE model of attitude-behavior processes. In S. Chaiken & Y. Trope (Eds.), *Dual-process theories in social psychology* (pp. 97–116). New York: Guilford Press.

Fiske, S. T., & Neuberg, S. L. (1990). A continuum of impression formation, from category-based to individuating processes: Influences of information and motivation on attention and interpretation. In M. P. Zanna (Ed.), *Advances in experimental social psychology* (Vol. 23, pp. 1–74). New York: Academic Press.

Gilbert, D. T., & Malone, P. S. (1995). The correspondence bias. *Psychological Bulletin, 117*, 21–38.

Greenwald, A. G., McGhee, D. E., & Schwartz, J. L. K. (1998). Measuring individual differences in implicit cognition: The implicit association test. *Journal of Personality and Social Psychology, 74*, 1464–1480.

Gregg, A. P., Seibt, B., & Banaji, M. R. (2006). Easier done than undone: Asymmetry in the malleability of implicit preferences. *Journal of Personality and Social Psychology, 90*, 1–20.

Haddock, G., Zanna, M. P., & Esses, V. M. (1993). Assessing the structure of prejudicial attitudes: The case of attitudes toward homosexuals. *Journal of Personality and Social Psychology, 65*, 1105–1118.

Higgins, E. T. (2006). Value from hedonic experience and engagement. *Psychological Review, 113*, 439–460.

Hofmann, W., Friese, M., & Strack, F. (2009). Impulse and self-control from a dual-systems perspective. *Perspectives on Psychological Science, 4*, 162–176.

Hofmann, W., Gschwendner, T., Friese, M., Wiers, R. W., & Schmitt, M. (2008). Working memory capacity and self-regulatory behavior: Toward an individual differences perspective on behavior determination by automatic versus controlled processes. *Journal of Personality and Social Psychology, 95,* 962–977.

Holmes, J. G. (1981). The exchange process in close relationships: Microbehavior and macromotives. In M. L. Lerner & S. Lerner (Eds.), *The justice motive in social behavior* (pp. 261–284). New York: Plenum.

Holmes, J. G. (2002). Interpersonal expectations as the building blocks of social cognition: An interdependence theory perspective. *Personal Relationships, 9,* 1–26.

Holmes, J. G., & Rempel, J. K. (1989). Trust in close relationships. In C. Hendrick (Ed.), *Review of personality and social psychology: Close relationships* (Vol. 10, pp. 187–219). Newbury Park, CA: Sage.

Karremans, J. C., & Aarts, H. (2007). The role of automaticity in determining the inclination to forgive close others. *Journal of Experimental Social Psychology, 43,* 902–917.

Kelley, H. H. (1979). *Personal relationships: Their structures and processes.* Hillsdale, NJ: Erlbaum.

Kunda, Z., & Spencer, S. J. (2003). When do stereotypes come to mind and when do they color judgment? A goal-based theoretical framework for stereotype activation and application. *Psychological Bulletin, 129,* 522–544.

LeBel, E. P., & Campbell, L. (2009). Implicit partner affect, relationship satisfaction, and the prediction of romantic breakup. *Journal of Experimental Social Psychology, 45,* 1291–1294.

Lee, S., Rogge, R. D., & Reis, H. T. (2010). Assessing the seeds of relationship decay: Using implicit evaluations to detect the early stages of disillusionment. *Psychological Science, 21,* 857–864.

Lydon, J. E., Menzies-Toman, D., Burton, K., & Bell, C. (2008). If-then contingences and the differential effects of the availability of an attractive alternative on relationship maintenance for men and women. *Journal of Personality and Social Psychology, 95,* 50–65.

Lykken, D. T., & Tellegen, A. (1993). Is human mating adventitious or the result of lawful choice? A twin study of mate selection. *Journal of Personality and Social Psychology, 65,* 56–68.

Macrae, C. N., & Johnston, L. (1998). Help, I need somebody: Automatic action and inaction. *Social Cognition, 16,* 400–417.

Mikulincer, M., Gillath, O., Halevy, V., Avihou, N., Avidan, S., & Eshkoli, N. (2001). Attachment theory and reactions to others' needs: Evidence that activation of the sense of attachment security promotes empathic responses. *Journal of Personality and Social Psychology, 81,* 1205–1224.

Mikulincer, M., Hirschberger, G., Nachmias, O., & Gillath, O. (2001). The affective components of the secure base schema: Affective priming with representations of attachment security. *Journal of Personality and Social Psychology, 81,* 305–321.

Mikulincer, M., & Shaver, P. R. (2001). Attachment theory and intergroup bias: Evidence that priming the secure base schema attenuates negative reactions to outgroups. *Journal of Personality and Social Psychology, 81,* 97–115.

Mikulincer, M., & Shaver, P. R. (2003). The attachment behavioral system in adulthood: Activation, psychodynamics, and interpersonal processes. In M. Zanna (Ed.), *Advances in experimental social psychology* (Vol. 35, pp. 52–153). New York: Academic Press.

Muraven, M., & Baumeister, R. F. (2000). Self-regulation and depletion of limited resources: Does self-control resemble a muscle? *Psychological Bulletin, 126,* 247–259.

Murray, S. L. (1999). The quest for conviction: Motivated cognition in romantic relationships. *Psychological Inquiry, 10,* 23–34.

Murray, S. L., Aloni, M., Holmes, J. G., Derrick, J. L., Stinson, D. A., & Leder, S. (2009). Fostering partner dependence as trust insurance: The implicit contingencies of the exchange script in close relationships. *Journal of Personality and Social Psychology, 96,* 324–348.

Murray, S. L., Bellavia, G., Rose, P., & Griffin, D. (2003). Once hurt, twice hurtful: How perceived regard regulates daily marital interactions. *Journal of Personality and Social Psychology, 84,* 126–147.

Murray, S. L., Derrick, J., Leder, S., & Holmes, J. G. (2008). Balancing connectedness and self-protection goals in close relationships: A levels of processing perspective on risk regulation. *Journal of Personality and Social Psychology, 94,* 429–459.

Murray, S. L., Griffin, D. W., Rose, P., & Bellavia, G. (2003). Calibrating the sociometer: The relational contingencies of self-esteem. *Journal of Personality and Social Psychology, 85,* 63–84.

Murray, S. L., & Holmes, J. G. (1993). Seeing virtues in faults: Negativity and the transformation of interpersonal narratives in close relationships. *Journal of Personalty and Social Psychology, 65,* 707–722.

Murray, S. L., & Holmes, J. G. (1997). A leap of faith? Positive illusions in romantic relationships. *Personality and Social Psychology Bulletin, 23,* 586–604.

Murray, S. L., & Holmes, J. G. (2009). The architecture of interdependent minds: A motivation-management theory of mutual responsiveness. *Psychological Review, 116,* 908–928.

Murray, S. L., & Holmes, J. G. (2011). *Interdependent minds: The dynamics of close relationships.* New York: Guilford Press.

Murray, S. L., Holmes, J. G., Aloni, M., Pinkus, R. T., Derrick, J. L., & Leder, S. (2009). Commitment insurance: Compensating for the autonomy costs of interdependence in close relationships. *Journal of Personality and Social Psychology, 97,* 256–278.

Murray, S. L., Holmes, J. G., & Collins, N. L. (2006). Optimizing assurance: The risk regulation system in relationships. *Psychological Bulletin, 132,* 641–666.

Murray, S. L., Holmes, J. G., & Griffin, D. W. (2000). Self-esteem and the quest for felt security: How perceived regard regulates attachment processes. *Journal of Personality and Social Psychology, 78,* 478–498.

Murray, S. L., Holmes, J. G., Griffin, D. W., Bellavia, G., & Rose, P. (2001). The mismeasure of love: How self-doubt contaminates relationship beliefs. *Personality and Social Psychology Bulletin, 27,* 423–436.

Murray, S. L., Holmes, J. G., MacDonald, G., & Ellsworth, P. (1998). Through the looking glass darkly? When self-doubts turn into relationship insecurities. *Journal of Personality and Social Psychology, 75,* 1459–1480.

Murray, S. L., Holmes, J. G., & Pinkus, R. T. (2010). A smart unconscious? Procedural origins of automatic partner attitudes in marriage. *Journal of Experimental Social Psychology, 46*, 650–656.

Murray, S. L., Pinkus, R. T., Harris, B., & Holmes, J. G., (2010). Unpublished data. State University of New York at Buffalo.

Murray, S. L., Pinkus, R. T., Holmes, J. G., Harris, B., Aloni, M., Gomillion, S., et al. (2010). *When rejection loses its motivational sting: The power of automatic partner attitudes*. Unpublished manuscript, University at Buffalo, The State University of New York.

Murray, S. L., Pinkus, R. T., Holmes, J. G., Harris, B., Gomillion, S., Aloni, M., Derrick, J., & Leder, S. (2011). Signaling when (and when not) to be cautious and self-protective: Impulsive and reflective trust in close relationships. *Journal of Personality and Social Psychology*, 101, 485–502.

Murray, S. L., Rose, P., Bellavia, G., Holmes, J., & Kusche, A. (2002). When rejection stings: How self-esteem constrains relationship-enhancement processes. *Journal of Personality and Social Psychology, 83*, 556–573.

Murray, S. L., Rose, P., Holmes, J. G., Derrick, J., Podchaski, E., Bellavia, G., et al. (2005). Putting the partner within reach: A dyadic perspective on felt security in close relationships. *Journal of Personality and Social Psychology, 88*, 327–347.

Olson, M. A., & Fazio, R. H. (2009). Implicit and explicit measures of attitudes: The perspective of the MODE model. In R. E. Petty, R. H. Fazio, & P. Brinol (Eds.), *Attitudes: Insights from the new implicit measures* (pp. 19–63). New York: Psychology Press.

Pierce, T., & Lydon, J. (1998). Priming relational schemas: Effects of contextually activated and chronically accessible interpersonal expectations on responses to a stressful event. *Journal of Personality and Social Psychology, 75*, 1441–1448.

Reis, H. T., Clark, M. S., & Holmes, J. G. (2004). Perceived partner responsiveness as an organizing construct in the study of intimacy and closeness. In D. Mashek & A. P. Aron (Eds.), *Handbook of closeness and intimacy* (pp. 201–225). Mahwah, NJ: Lawrence Erlbaum.

Rusbult, C. E., & Van Lange, P. A. M. (2003). Interdependence, interaction, and relationships. *Annual Review of Psychology, 54*, 351–375.

Scinta, A., & Gable, S. L. (2007). Automatic and self-reported attitudes in romantic relationships. *Personality and Social Psychology Bulletin, 33*, 1008–1022.

Sprecher, S., Schmeeckle, M., & Felmless, D. (2006). The principle of least interest: Inequality in emotional invollvement in romantic relationships. *Journal of Family Issues, 27*, 1–26.

Waller, W. (1938). *The family: A dynamic interpretation*. New York: Cordon.

Wegner, D. M. (2002). *The illusion of conscious will*. Cambridge, MA: MIT Press.

Wieselquist, J., Rusbult, C. E., Foster, C. A., & Agnew, C. R. (1999). Commitment, pro-relationship behavior, and trust in close relationships. *Journal of Personality and Social Psychology, 77*, 942–966.

Wilson, T. D. (2002). *Strangers to ourselves: Discovering the adaptive unconscious*. Cambridge, MA: Harvard University Press.

Wilson, T. D., Lindsey, S., & Schooler, T. Y. (2000). A dual model of attitudes. *Psychological Review, 107*, 101–126.

Zayas, V., & Shoda, Y. (2005). Do automatic reactions elicited by thoughts of romantic partner, mother, and self relate to adult romantic attachment? *Personality and Social Psychology Bulletin, 8*, 1011–1025.

CHAPTER 9

Eustress in Romantic Relationships

TIMOTHY J. LOVING

The University of Texas at Austin

BRITTANY L. WRIGHT

The University of Texas at Austin

> If psychological events can produce stress effects in the body, it should be in principle possible for psychological events to have beneficial effects. We have paid far too little attention to the possibilities of such salubrious influences. Although many persons are prepared to think of psychological stress as impairing health, we are less well prepared to think about positive states of mind having beneficial effects.
>
> **Lovallo (2005, p. 233)**

One of the more exciting advances over the past 20 years in the study of close relationships is the increased focus on the link between romantic relationships and health. The vast majority of this work has made use of relatively simple methods to assess physiological function in an attempt to elucidate how romantic relationship processes influence underlying health-relevant physiological outcomes. The collective scope of these investigations is impressive. We now know a great deal about how close relationships affect our bodies, including but not limited to cardiovascular (Holt-Lunstad, Uchino, Smith, Olson-Cerny, & Nealey-Moore,

2003; Nealey-Moore, Smith, Uchino, Hawkins, & Olson-Cerny, 2007), neuroendocrine (Gouin et al., 2010; Grewen, Girdler, Amico, & Light, 2005; Loving, Heffner, Kiecolt-Glaser, Glaser, & Malarkey, 2004), and immune outcomes (Dopp, Miller, Myers, & Fahey, 2000; Kiecolt-Glaser, Malarkey, Chee, & Newton, 1993; Miller, Dopp, Myers, Stevens, & Fahey, 1999) as well as "objective" markers of physical health (e.g., wound healing; Kiecolt-Glaser et al., 2005) and neural activity (Aron et al., 2005; Bartels & Zeki, 2000). This growing body of research is paralleled by what was (and, in our view, remains to be) a needed expansion of the contexts in which romantic relationships have been studied, including conflict (Powers, Pietromonaco, Gunlicks, & Sayer, 2006; Robles, Shaffer, Malarkey, & Kiecolt-Glaser, 2006), support (Kiecolt-Glaser et al., 2005), playful or comforting interactions (Gonzaga, Turner, Keltner, Campos, & Altemus, 2006; Holt-Lunstad, Birmingham, & Light, 2008), temporary and permanent partner separations (Diamond, Hicks, & Otter-Henderson, 2008; Fisher, Brown, Aron, Strong, & Mashek, 2010; Sbarra, Law, Lee, & Mason, 2009), and an increased focus on nonmarital romances (e.g., Loving, Gleason, & Pope, 2009; Powers et al., 2006). This progress in the study of mind–body connections within the context of romantic relationships is illustrated nicely by the first chapter dedicated solely to the physiology of interpersonal relationships in the most recent *Cambridge Handbook of Personal Relationships* (Loving, Heffner, & Kiecolt-Glaser, 2006) as well as special issues dedicated to the topic in the two journals of the International Association for Relationship Research—the *Journal of Social and Personal Relationships* (Floyd, 2006) and *Personal Relationships* (2011).

Our objective with this chapter is to address what we view to be one of the gaps in our current knowledge about the physiological consequences associated with romantic relationships. Put simply, borrowing the "for better or for worse" phrase oft-cited in wedding vows, we know far more about the *worse* than we do the *better* (Fincham & Beach, 2010). Specifically, the potentially harmful or negative physiological consequences of romantic relationships dominate our understanding of how relationship processes affect the body. Substantially less is known about how subjectively or affectively positive events and contexts influence physiological parameters.

We suggest that the continued melding of the fields of close relationships and psychophysiology provides a unique opportunity to (re) introduce an understudied concept and phenomenon, if not a forgotten one, in the general stress literature: eustress (Selye, 1978b). To accomplish this objective, we first distinguish between *dis*tress and *eu*stress and briefly discuss why we believe the focus, both in and out of the romantic relationship context, has been on distress. Next, we argue for the importance of studying eustress-inducing relationship contexts and highlight recent work from our lab on the physiological consequences of positive romantic relationship transitions (i.e., falling in love). Finally, we conclude with some preliminary thoughts on how studying romantic relationships has the potential to expand our understanding of the stress–health link more broadly.

Distress Versus Eustress

Scientific inquiry into the link between romantic relationships and physiology often focuses on "stressful" relationship events, such as conflict, dissolution, and temporary separations. Although we recognize such contexts are not universally negative (e.g., Lewandowski & Bizzoco, 2007), the term *stressful* inherently carries a negative connotation. In fact, Wikipedia defines stress as "the consequence *of the failure of an organism*—human or animal—to respond appropriately to emotional or physical threats, whether actual or imagined" (Stress [biology], 2010; italics added). Interestingly, Hans Selye, who coined the term stress (Selye, 1975), later explained that his reference to a "nonspecific response of the body to any demand" (i.e., the stress response; Selye, 1978b, p. 74) is not reserved solely for an organism's reaction to noxious stimuli. Rather, Selye distinguished between *distress* and *eustress*, or bad versus good (i.e., euphoric) stress, respectively (Selye, 1978b). Many operationalizations and definitions of eustress have been provided, including defining eustress as individual differences in the ability to view negative events as positive (Selye, 1978a), and the optimal level of arousal needed to maximize performance on a task (Gibbons, Dempster, & Moutray, 2008). Our focus in this chapter is on seemingly *positive* events that produce the tell-tale nonspecific stress response, including activation of the hypothalamic-pituitary-adrenal (HPA) axis, resulting in the release of cortisol (for a brief overview of the stress response and biological markers, see Loving et al., 2006). This set of criteria is most consistent with Edwards and Cooper (1988), who defined eustress as "a positive discrepancy between an individual's perceived state and desired state, provided that the presence of this discrepancy is considered important by the individual" (p. 1448). Likewise, distress can be defined as a negative discrepancy between an individual's perceived state and desired state, provided that the presence of this discrepancy is considered important by the individual.

The key to these definitions of distress and eustress is the focus on discrepancies, whether positive or negative, which necessitate adaptation on the part of the individual. We adopt this definition because of the central role close relationships play in individuals' lives (i.e., relationships are important to us; Kelley, 1979). Further, the development of romantic relationships requires transitioning into more (versus less) desired states depending on individuals' perceptions and overall goals (e.g., falling in love, becoming more committed, deciding to cohabit). Thus, romantic relationships offer multiple possibilities for the creation of positive (and negative) discrepancies between individuals' current and desired states.

As we noted already, despite the introduction of the eustress concept over 3 decades ago, distress dominates the stress-and-health literature (Nelson & Cooper, 2005; Nelson & Simmons, 2003). Part of the reason for the empirical focus on distress is that early attempts to predict health outcomes from measures of stress suggested that distress affects health more than does eustress. For example, Holmes and Rahe's (1967) classic Social Readjustment Rating Scale (SRRS), a life-events checklist, was developed under the supposition that positive and negative events, both of which require adaptation on the part of the individual, induce stress (i.e., change = stress). This measurement approach was quite consistent with Selye's view of stress but was nonetheless critiqued on several grounds, including concerns that the SRRS confounded positive and negative events even though only negative events proved predictive of poor mental health (e.g., Zautra & Reich, 1983). Subsequent self-report measures of stress focused on minor life events but still preserved the idea of negative (i.e., hassles) and positive (i.e., uplifts) stressors, but, with few exceptions, hassles were associated more strongly with health outcomes than were uplifts. When uplifts do predict health outcomes, the association is generally accounted for by shared variance between hassles and uplifts (Delongis, Coyne, Dakof, Folkman, & Lazarus, 1982). In a few studies, decreases in uplifts predicted illness onset (Evans & Edgerton, 1991; Evans, Pitts, & Smith, 1988); no positive effects, however, have been found for a higher number of uplifts. Given the significant impact negative events relative to positive events have on health, it is no surprise researchers have slowly shifted focus away from more positive aspects of life.

The concept of eustress is also wrought with methodological and conceptual challenges. For example, one reason positive life events, or uplifts, do not predict health outcomes is because the focus is often on negative health outcomes or other signs of distress. Some have argued that the importance of positive life events is perhaps best illustrated when assessing health outcomes that adopt less of a biomedical "health is the absence of disease" approach and instead consider wellness on a continuum (Edwards & Cooper, 1988). Additionally, researchers often identify an event as eustressful if participants, after the event, report positive psychological states (e.g., McGowan, Gardner, & Fletcher, 2006; Nelson & Simmons, 2003). The problem with this approach, however, is the precipitating event is generally one that is objectively noxious (e.g., work stress, role conflict), leaving open the possibility that what is being tapped in these situations is individual coping styles or some other individual difference variable. For example, a line of work by Tomaka, Blascovich, and colleagues indicates that, when faced with a potentially stressful event, individuals appraise the situation as being "threatening" or "challenging." Individuals perceive events as threatening when the events are self-relevant and the individuals do not have adequate resources (e.g., skills, knowledge) to cope with the demands (e.g., physical or psychological requirements) of the situation; when individuals have the resources to meet the demands of a particular event, it is perceived as challenging (e.g., Tomaka, Blascovich, Kelsey, & Leitten, 1993). These different motivational states elicit different psychological and physiological responses when confronting stressful situations (Tomaka, Palacios, Schneider, Colotla, Concha, & Herrald, 1999). For example, *challenges*, relative to threats, elicit positive affect (Maier, Waldstein, & Synowski, 2003; Tomaka et al., 1999), better perceived and actual performance on a task (Blascovich, 2004; Drach-Zahavy & Erez, 2002), higher task engagement (Maier et al., 2003), and attenuated reports of pain (e.g., cold pressor task; Tomaka, Blascovich, Kibler, & Ernst, 1997). Notably, the positive reactions to challenges are often coupled with an *increase* in cardiovascular activity and low systemic vascular resistance (Tomaka et al., 1993; see also McGowan et al., 2006, who link the concept of eustress to challenge appraisals).

In short, we believe that to best understand how eustress affects physical well-being researchers need to study individual physiological outcomes in the context of events that are important to the individual and create a "positive discrepancy" that evokes a "nonspecific stress response." Identification of such events has been challenging, although not entirely elusive. For example, in a recent study, Rietveld and van Beest (2007) examined fluctuations of a host of physiological outcomes (e.g., shortness of breath, lung function, heart rate, blood pressure, dizziness, nausea, and heart pounding) at baseline and before and after participants rode a roller coaster. Although participants rated the roller-coaster ride positively, blood pressure, heart rate, and somatic symptoms (e.g., heart pounding) were highest before or after the ride relative to baseline. In other words, this study demonstrates that physiological outcomes that are often perceived as distressful can be associated with a positive state of mind. Such a stress context is consistent with Selye's view that stress results in a nonspecific physiological response but that response can be provoked by positive and negative stimuli. The study by Rietveld and van Beest suggests that positively arousing situations can induce such a response. A few other studies have revealed similar effects, particularly when the manipulation used to induce the positive arousal is "self-relevant and engaging" (Pressman & Cohen, 2005), but in most cases these other studies simply prime positive moods or emotions rather than look at naturally occurring reactions to positive stressors.

Relationships Are Like a Roller Coaster

We believe that developing romantic relationships, and the transitions that define them, provide an ideal context with which to study eustress. Although all available evidence suggests that transitions into romantic relationships are associated with positive outcomes (Lee & Gramotnev, 2007), the impact of these transitions on well-being have received scant empirical attention. Yet events such as "starting a love relationship" and "begin to date" are considered *positive* life events in the health literature (Reich & Zautra, 1981), and any negative effects of relationship dissolution are diminished if not reversed when individuals begin a new relationship (Wheaton, 1990). This finding should come as no surprise. The need to belong is a *fundamental* need (Baumeister & Leary, 1995), and fulfillment of that need, or movement to a more desired state, should procure benefits. Some of the strongest evidence for this assertion comes from recent work on the concept of early stage passionate love.

The Physiology of Passionate Love

The process of falling in love, which is characterized by high levels of passion, is a key component of developing romantic relationships. Passionate love refers to the subjective experience characteristic of newly developed romances resulting in physiological, psychological, and behavioral outcomes. Given its apparent universality across time and cultures, passionate love is hypothesized to have biological substrates that evolved as a primary aspect of the human mating system (Carter, 1998) and yields multiple benefits for individuals (Stefano & Esch, 2005).

The experience of passionate love activates dopamine-rich sections of the brain known to be associated with the *reward-motivation system* (Aron et al., 2005; Bartels & Zeki, 2000). Aron and colleagues compared the brain activation of individuals while they viewed a picture of their (nonmarital) romantic partner with brain activation while they viewed the picture of an opposite-sex friend with whom the participants had no romantic relationship. Comparison of the relative patterns of brain activation indicated stronger activity in the ventral tegmental and anteromedial caudate areas when participants viewed their partners. These same areas of the brain are activated when individuals are experiencing other rewarding states, which is why they are labeled the *pleasure center* of the brain (Wise & Rompre, 1989).

Passionate love is associated with changes in circulating levels of neurotrophins and stress hormones. Individuals who score high on a self-report measure of passionate love show increased levels of the neurotrophin nerve growth factor (NGF) relative to the unattached or those in long-term established romances (Emanuele, Politi, Bianchi, Minoretti, Bertona, & Geroldi,

2006). Further, blood samples drawn from individuals experiencing passionate love reveal significantly greater circulating levels of cortisol relative to control groups (Marazziti & Canale, 2004). The differences (NGF and cortisol) are not detectable when participants are retested after a year or more. These physiological findings suggest that the novelty of developing relationships dissipates within a year, which is consistent with theorizing that passion peaks early in the development of romantic relationships (Baumeister & Bratslavsky, 1999) and then declines over time (Tucker & Aron, 1993). Thus, passionate love is an emotionally intense phase of developing relationships characterized by high levels of physiological arousal and intense positive mood.

Passionate Love and Acute Stress Responses

As we noted already, passionate love appears to result in increased circulation of cortisol, a key hormone involved in the stress response that is secreted via activation of the HPA axis (Lovallo, 2005; Selye, 1978b). Past work on the neurobiology of passionate love is built on the assumption that the number of hours individuals spend thinking about their partners is an important component of the passionate love experience (Emanuele et al., 2006; Marazziti & Canale, 2004). In a recent study, we used a validated measure of relationship-focused cognition to determine whether individual differences in how much people think about their partners and romances alter HPA-axis activity (Loving, Crockett, & Paxson, 2008). Given passionate love's effects on neural processes and attentiveness to environmental stimuli (Aron et al., 2005; Bianchi-Demicheli, Grafton, & Ortigue, 2006), we reasoned that individuals who demonstrate a tendency to think more about their relationships would be particularly susceptible to increased cortisol levels when in the throes of a passionate relationship.

To test this hypothesis, we assessed acute shifts in salivary cortisol levels after participants ($n = 29$ women; average relationship duration was just over 4 months), all of whom were experiencing passionate love, were asked to either reflect on their romantic partner or an opposite-sex friend (condition determined via random assignment). All participants, regardless of condition, first underwent a brief relaxation exercise, after which a baseline saliva sample was collected. Next, women in the partner condition underwent a guided imagery protocol in which they were asked to think about and reflect on falling in love with their partner. Specifically, they were instructed* to:

* What follows is an abbreviated version of the guided imagery protocol. Some text and conversational pauses have been omitted. Additionally, one female participant was involved in a same-sex relationship; all pronouns in the script were modified accordingly, and this participant brought in a picture of a same-sex friend to serve as a control. Contact the first author for the full guided imagery script.

> Picture (partner's) face and try to visualize all the details about him. Think about the first time you saw or met (partner) and how he made you feel. Picture where you were and what each of you was wearing. Really try to capture and understand the feelings you were having during this time. What thoughts went through your mind when you first saw (partner)? Think about the first time the two of you did something as a couple. What did you both do? Think about the first time you realized you were in love with (partner). Go ahead and take a moment to recreate this memory as vividly and fully as you can.... Think about the times the two of you have laughed together and how you feel when he is close to you. What things do you like most about (partner)? The key thing is that all you are thinking about is (partner) and all your feelings for him are being felt by you right now.

Following the guided imagery manipulation, participants were instructed to talk (3 minutes) and then write (10 minutes) about "all the things going through your head." Participants subsequently provided three additional saliva samples over the course of the following 30 minutes. Participants assigned to the friend condition underwent an identical procedure, except their guided imagery session focused on becoming friends with an individual of the same sex and age as their romantic partner, but somebody with whom they had not had or with whom they did not plan to have a romantic relationship. The use of an opposite-sex friend served as a control condition and follows the method employed in recent functional magnetic resonance imaging (fMRI) research on passionate love (Aron et al., 2005). To assess relationship-focused cognition, all participants completed Cate and colleagues' (1995) relationship thinking scale (e.g., "I find myself at times drifting off and thinking about my relationship with my partner"; 1 = extremely unlike me; 5 = extremely like me).

The pattern of results indicated that cortisol production can be increased or decreased via mere reflection of the falling-in-love process. Specifically, women who engaged in more relationship-focused thinking experienced acute increases in cortisol after simply being asked to reflect on their romantic relationship and partner. Interestingly, women with a lesser tendency to engage in relationship-focused thinking (i.e., lower scores on the relationship thinking scale) also showed an initial rise in cortisol after reflecting on their partners, but the effect diminished shortly after our experimental manipulation concluded.

Passionate Love: Distress or Eustress?

We believe the increased cortisol response previously noted reflects eustress. In addition to the biological consequences already outlined, passionate love increases the prevalence of positive emotions (Kim & Hatfield,

2004). Individuals in passionate love focus on positive emotions (Brand, Luethi, von Planta, Hatzinger, & Holsboer-Trachsler, 2007) and more quickly recognize positive feeling words when primed with their partners' names (Bianchi-Demicheli, Grafton, & Ortigue, 2006). Further, our manipulation likely induces a positive state of arousal, which as we noted already has been shown in other studies to increase cortisol (Pressman & Cohen, 2005). In other words, the process of falling in love, or becoming more involved with another, creates a positive discrepancy between an individual's perceived state and desired state, and we believe there is little doubt that the presence of this discrepancy is considered important by the individual (Aron et al., 2005; Kelley, 1979).

Furthermore, we suggest that the eustress associated with the experience of passionate love likely results in salubrious consequences. We believe this to be the case for several reasons. First, although becoming more intimately involved with another requires adaption, and adaptation challenges an organism, the positive emotions that characterize the falling-in-love transition may actually protect individuals from negative physical and mental health outcomes by facilitating adaptation to stressful situations (Cohen & Hoberman, 1983; Folkman & Moskowitz, 2007). Further, life events individuals perceive as being "good" can have a beneficial effect on individual outcomes, including physical health (Snyder, Roghmann, & Sigal, 1993), although, as we reviewed previously, such evidence is limited.

Interestingly, a recent meta-analysis on the link between positive affect and physical health suggests that intense bouts of state-positive affect may undermine health, particularly for those with an underlying condition (e.g., asthma), but this conclusion requires several qualifications (for a review, see Pressman & Cohen, 2005). In contrast, Stefano, Stefano, and Esch (2008) argued that the stress responses observed in individuals experiencing love, more generally, reflect an adaptive anticipatory stress response that ultimately benefits physiological functioning and health. Such an idea is not unwarranted. Others have documented the immediate benefits of acute stress for immune system function (e.g., Dhabhar, Saul, Daugherty, Holmes, Bouley, & Oberyszyn, 2010; Dhabhar & Viswanathan, 2005), although to our knowledge none have done so in the context of experimentally induced eustress. Further, it is possible that passionate love results in more chronic basal increases in cortisol (Marazziti & Canale, 2004). Although we can find no evidence that there is any objective truth to the notion of being lovesick, we cannot rule out such a possibility in light of the potential chronic increases in cortisol during the falling-in-love transition. In short, although we are hesitant to state that the physiological responses we observe in those falling in love necessarily promote health, we believe there is reason to believe such an outcome is possible, particularly given the salubrious effects of positive affect on health-relevant physiological parameters when assessed in naturalistic contexts (Costanzo et al., 2004; Stone, Cox, Valdimarsdottir, Jandorf, & Neale, 1987).

Importantly, within the context of nonmarital romances, both changes in the state (e.g., casual vs. exclusive) and fate (e.g., intact vs. dissolved) of a romance require individuals to adapt affectively (Nieder & Sieffge-Krenke, 2001; Sbarra, 2006), behaviorally (Kelley & Thibaut, 1978), and cognitively (Agnew, Van Lange, Rusbult, & Langston, 1998; Aron, Paris, & Aron, 1995). As a result, we do not believe the experience of passionate love is the only eustress-inducing event or transition in romantic relationships. Essentially, any experienced change in relationship status that is desired by the individual holds potential to induce a stress response. For example, we have also documented increased HPA-axis activity in long-term dating partners who are asked to talk about the idea of someday marrying each other. This effect was moderated by the extent to which couple members had not previously talked about or thought about getting married. Importantly, we could find no evidence that this conversation was perceived as distressful or otherwise negative by participants (Loving et al., 2009). Rather, informal information gathered during study debriefing suggested that participants were universally pleased to have had the conversation (regardless of the outcome). Although we cannot rule out the possibility that participants found the conversation threatening, we believe the results, in light of the results from our passionate love study, are consistent with the idea that reflection on a relationship transition is stressful, but such stress may be indicative of eustress rather than distress.

The Benefits of Studying Eustress in Romantic Relationships

There are at least three benefits of studying eustress in the romantic relationship context. First, as we noted already, the life course of romantic relationships is replete with potential eustress-inducing transitions. As a result, studying stress responses, broadly, in romantic relationships can help to further understanding of how the stress response generally influences health by promoting understanding of how both good and bad stress ultimately contribute to overall well-being versus the current lopsided focus on distress (Nelson & Simmons, 2003).

Second, the results from our laboratory-based studies suggest that the physiological and health effects of romantic relationships can be studied in and out of the lab. In light of the apparent disconnect in findings regarding the health consequences of positive affect between lab-based and naturalistic studies (Pressman & Cohen, 2005), which typically focus on vastly different contexts, we believe this to be an important methodological bonanza. Romantic relationship transitions obviously occur in the real world, can last for significant periods of time, and are highly amenable to diary and other longitudinal designs (e.g., Aron et al., 2005; Fine & Sacher, 1997; Marazziti & Canale, 2004; Sprecher, Felmlee, Metts, Fehr, & Vanni, 1998). But, as we have demonstrated, simply priming individuals to think about transitions can be an equally powerful methodological tool. As a result, the study of mind–body connections within the context of romantic relationships promises to inform understanding of more basic research into the effects of positive mental states on physical health by allowing eustress- (and distress-) inducing transitions to be studied both in and out of the laboratory.

Finally, given that eustress-inducing relationship transitions are amenable to priming, their use also permits researchers to assess how individuals currently experiencing a positive relationship event are able to recover from non-relationship-based stressors. Some have argued that one way eustress benefits health is via its ability to affect how individuals cope with or appraise more distressing life events (Milsum, 1985; Nelson & Cooper, 2005; Selye, 1978a). Thus, priming relationship transitions, given their known impact on stress responses, provides researchers an opportunity to assess how both eustress-induced physiological responses and psychological responses (e.g., positive affect, coping resources) affect individuals' acute stress reactivity and recovery, both of which have important implications for physical health outcomes (Forcier et al., 2006; Ong, Bergeman, Bisconti, & Wallace, 2006).

Summary

It has become almost cliché to state that romantic relationships "get under the skin." This is unfortunate given that the science of the couple with regard to the mind–body connection in personal relationships still has significant room for growth, particularly in light of the focus on negative or distressful relationship phenomena (e.g., conflict). Importantly, the layperson often talks about the "ups and downs" of relationships, yet assessment of the ups of relationships in terms of how they impact couple physiology is severely lacking. We believe the ups and downs that characterize romantic relationship transitions provide a unique methodological opportunity for researchers to advance understanding of how bad stressors (i.e., distress) and good stressors (i.e., eustress) contribute to overall well-being. To date, the experience of falling in love, or passionate love, is the one positive relationship stressor that has received empirical attention, and the physiological consequences of this eustressor mimic what is observed in individuals experiencing negative stressors. The objective health outcomes of these eustress-induced outcomes remain to be seen, but such data will go a long way toward promoting a complete understanding of why and how relationships truly do get under the skin *for better or for worse.*

References

Agnew, C. R., Van Lange, P. A. M., Rusbult, C. E., & Langston, C. A. (1998). Cognitive interdependence: Commitment and the mental representation of close relationships. *Journal of Personality and Social Psychology, 74*(4), 939–954.

Aron, A., Fisher, H., Mashek, D. J., Strong, G., Li, H., & Brown, L. L. (2005). Reward, motivation, and emotion systems associated with early-stage intense romantic love. *Journal of Neurophysiology, 94*(1), 327–337.

Aron, A., Paris, M., & Aron, E. N. (1995). Falling in love: Prospective studies of self-concept change. *Journal of Personality and Social Psychology, 69*(6), 1102–1112.

Bartels, A., & Zeki, S. (2000). The neural basis of romantic love. *NeuroReport, 11*(17), 3829–3834.

Baumeister, R. F., & Bratslavsky, E. (1999). Passion, intimacy, and time: Passionate love as a function of change in intimacy. *Personality and Social Psychology Review, 3*(1), 49–67.

Baumeister, R. F., & Leary, M. R. (1995). The need to belong: Desire for interpersonal attachments as a fundamental human motivation. *Psychological Bulletin, 117*(3), 497–529.

Bianchi-Demicheli, F., Grafton, S. T., & Ortigue, S. (2006). The power of love on the human brain. *Social Neuroscience, 1*(2), 90–103.

Blascovich, J., Seery, M. D., Mugridge, C. A., Norris, R. K., & Weisbuch, M. (2004). Predicting athletic performance from cardiovascular indexes of challenge and threat. *Journal of Experimental Social Psychology, 40*(5), 683–688. doi: 10.1016/j.jesp.2003.10.007

Brand, S., Luethi, M., von Planta, A., Hatzinger, M., & Holsboer-Trachsler, E. (2007). Romantic love, hypomania, and sleep pattern in adolescents. *Journal of Adolescent Health, 41*(1), 69–76.

Carter, S. C. (1998). Neuroendocrine perspectives on social attachment and love. *Psychoneuroendocrinology, 23*(8), 779–818.

Cate, R. M., Koval, J., Lloyd, S. A., & Wilson, G. (1995). Assessment of relationship thinking in dating relationships. *Personal Relationships, 2*(2), 77–95.

Cohen, S., & Hoberman, H. M. (1983). Positive events and social supports as buffers of life change stress. *Journal of Applied Social Psychology, 13*(2), 99–125.

Costanzo, E. S., Lutgendorf, S. K., Kohut, M. L., Nisly, N., Rozeboom, K., Spooner, S., et al. (2004). Mood and cytokine response to influenza virus in older adults. *Journals of Gerontology. Series A, Biological Sciences and Medical Sciences, 59*(12), 1328–1333.

Delongis, A., Coyne, J. C., Dakof, G., Folkman, S., & Lazarus, R. S. (1982). Relationship of daily hassles, uplifts, and major life events to health status. *Health Psychology, 1*(2), 119–136.

Dhabhar, F. S., Saul, A. N., Daugherty, C., Holmes, T. H., Bouley, D. M., & Oberyszyn, T. M. (2010). Short-term stress enhances cellular immunity and increases early resistance to squamous cell carcinoma. *Brain, Behavior, and Immunity, 24*(1), 127–137.

Dhabhar, F. S., & Viswanathan, K. (2005). Short-term stress experienced at time of immunization induces a long-lasting increase in immunologic memory. *American Journal of Physiology: Regulatory, Integrative, and Comparative Physiology, 289*(3), R738–R744.

Diamond, L. M., Hicks, A. M., & Otter-Henderson, K. D. (2008). Every time you go away: Changes in affect, behavior, and physiology associated with travel-related separations from romantic partners. *Journal of Personality and Social Psychology, 95*(2), 385–403.

Dopp, J. M., Miller, G. E., Myers, H. F., & Fahey, J. L. (2000). Increased natural killer-cell mobilization and cytotoxicity during marital conflict. *Brain, Behavior, and Immunity, 14*(1), 10–26.

Drach-Zahavy, A., & Erez, M. (2002). Challenge versus threat effects on the goal-performance relationship. *Organizational Behavior and Human Decision Processes, 88*(2), 667–682.

Edwards, J. R., & Cooper, C. L. (1988). The impacts of positive psychological states on physical health: A review and theoretical framework. *Social Science & Medicine, 27*(12), 1447–1459.

Emanuele, E., Politi, P., Bianchi, M., Minoretti, P., Bertona, M., & Geroldi, D. (2006). Raised plasma nerve growth factor levels associated with early-stage romantic love. *Psychoneuroendocrinology, 31*, 288–294.

Evans, P. D., & Edgerton, N. (1991). Life-events and mood as predictors of the common cold. *British Journal of Medical Psychology, 64*(Pt 1), 35–44.

Evans, P. D., Pitts, M. K., & Smith, K. (1988). Minor infection, minor life events and the four day desirability dip. *Journal of Psychosomatic Research, 32*(4–5), 533–539.

Fincham, F. D., & Beach, S. R. H. (2010). Of memes and marriage: Toward a positive relationship science. *Journal of Family Theory & Review, 2*, 4–24.

Fine, M. A., & Sacher, J. A. (1997). Predictors of distress following relationship termination among dating couples. *Journal of Social & Clinical Psychology, 16*(4), 381–388.

Fisher, H. E., Brown, L. L., Aron, A., Strong, G., & Mashek, D. (2010). Reward, addiction, and emotion regulation systems associated with rejection in love. *Journal of Neurophysiology, 104*(1), 51–60.

Floyd, K. (2006). Physiology and human relationships [Special issue]. *Journal of Social and Personal Relationships, 23*(2).

Folkman, S., & Moskowitz, J. T. (2007). Positive affect and meaning-focused coping during significant psychological stress. In M. Hewstone, H. A. W. Schut, J. B. F. De Wit, & K. Van Den Bos (Eds.), *The scope of social psychology: Theory and applications* (pp. 193–208). New York: Psychology Press.

Forcier, K., Stroud, L. R., Papandonatos, G. D., Hitsman, B., Reiches, M., Krishnamoorthy, J., et al. (2006). Links between physical fitness and cardiovascular reactivity and recovery to psychological stressors: A meta-analysis. *Health Psychology, 25*(6), 723–739.

Gibbons, C., Dempster, M., & Moutray, M. (2008). Stress and eustress in nursing students. *Journal of Advanced Nursing, 61*(3), 282–290.

Gonzaga, G. C., Turner, R. A., Keltner, D., Campos, B., & Altemus, M. (2006). Romantic love and sexual desire in close relationships. *Emotion, 6*(2), 163–179.

Gouin, J. P., Carter, C. S., Pournajafi-Nazarloo, H., Glaser, R., Malarkey, W. B., Loving, T. J., et al. (2010). Marital behavior, oxytocin, vasopressin, and wound healing. *Psychoneuroendocrinology, 35,* 1082–1090.

Grewen, K. M., Girdler, S. S., Amico, J., & Light, K. C. (2005). Effects of partner support on resting oxytocin, cortisol, norepinephrine, and blood pressure before and after warm partner contact. *Psychosomatic Medicine, 67*(4), 531–538.

Holmes, T. H., & Rahe, R. H. (1967). The social readjustment rating scale. *Journal of Psychosomatic Research, 11*(2), 213–218.

Holt-Lunstad, J., Birmingham, W. A., & Light, K. C. (2008). Influence of a "warm touch" support enhancement intervention among married couples on ambulatory blood pressure, oxytocin, alpha amylase, and cortisol. *Psychosomatic Medicine, 70*(9), 976–985.

Holt-Lunstad, J., Uchino, B. N., Smith, T. W., Olson-Cerny, C., & Nealey-Moore, J. B. (2003). Social relationships and ambulatory blood pressure: Structural and qualitative predictors of cardiovascular function during everyday social interactions. *Health Psychology, 22*(4), 388–397.

Kelley, H. H. (1979). *Personal relationships: Their structures and processes.* Hillsdale, NJ: Lawrence Erlbaum Associates.

Kelley, H. H., & Thibaut, J. W. (1978). *Interpersonal relations: A theory of interdependence.* New York: Wiley-Interscience.

Kiecolt-Glaser, J. K., Loving, T. J., Stowell, J. R., Malarkey, W. B., Lemeshow, S., Dickinson, S. L. et al. (2005). Hostile marital interactions, proinflammatory cytokine production, and wound healing. *Archives of General Psychiatry, 62*(12), 1377–1384.

Kiecolt-Glaser, J. K., Malarkey, W. B., Chee, M., Newton, T., Cacioppo, J. T., Mao, H. Y., et al. (1993). Negative behavior during marital conflict is associated with immunological down-regulation. *Psychosomatic Medicine, 55*(5), 395–409.

Kim, J., & Hatfield, E. (2004). Love types and subjective well-being: A cross-cultural study. *Social Behavior and Personality, 32*(2), 173–182.

Lee, C., & Gramotnev, H. (2007). Life transitions and mental health in a national cohort of young Australian women. *Developmental Psychology, 43*(4), 877–888.

Lewandowski, G. W., Jr., & Bizzoco, N. M. (2007). Addition through subtraction: Growth following the dissolution of a low quality relationship. *Journal of Positive Psychology, 2*(1), 40–54.

Lovallo, W. R. (2005). *Stress & health: Biological and psychological interactions* (2nd ed.). Thousand Oaks, CA: Sage Publications, Inc.

Loving, T. J., & Campbell, L. (2011). Mind–body connections in personal relationships: What close relationships have to offer. *Personal Relationships,* 18, 165–169.

Loving, T. J., Crockett, E. E., & Paxson, A. A. (2008). *Passionate love and relationship thinkers: Experimental evidence for acute cortisol elevations in women.* Manuscript under review.

Loving, T. J., Gleason, M. E. J., & Pope, M. T. (2009). Transition novelty moderates daters' cortisol responses when talking about marriage. *Personal Relationships, 16*(2), 187–203.

Loving, T. J., Heffner, K. L., & Kiecolt-Glaser, J. K. (2006). Physiology and interpersonal relationships. In A. Vangelisti & D. Perlman (Eds.), *Cambridge handbook of personal relationships* (pp. 385–405). New York: Cambridge University Press.

Loving, T. J., Heffner, K. L., Kiecolt-Glaser, J. K., Glaser, R., & Malarkey, W. B. (2004). Stress hormone changes and marital conflict: Spouses' relative power makes a difference. *Journal of Marriage and Family, 66*(3), 595–612.

Maier, K., Waldstein, S., & Synowski, S. (2003). Relation of cognitive appraisal to cardiovascular reactivity, affect, and task engagement. *Annals of Behavioral Medicine, 26*(1), 32–41.

Marazziti, D., & Canale, D. (2004). Hormonal changes when falling in love. *Psychoneuroendocrinology, 29*, 931–936.

McGowan, J., Gardner, D., & Fletcher, R. (2006). Positive and negative affective outcomes of occupational stress. *New Zealand Journal of Psychology, 35*(2), 92–98.

Miller, G. E., Dopp, J. M., Myers, H. F., Stevens, S. Y., & Fahey, J. L. (1999). Psychosocial predictors of natural killer cell mobilization during marital conflict. *Health Psychology, 18*(3), 262–271.

Milsum, J. H. (1985). A model of the eustress system for health/illness. *Behavioral Science, 30*(4), 179–186.

Nealey-Moore, J. B., Smith, T. W., Uchino, B. N., Hawkins, M. W., & Olson-Cerny, C. (2007). Cardiovascular reactivity during positive and negative marital interactions. *Journal of Behavioral Medicine, 30*(6), 505–519.

Nelson, D. L., & Cooper, C. (2005). Guest editorial: Stress and health: A positive direction. *Stress and Health: Journal of the International Society for the Investigation of Stress, 21*(2), 73–75.

Nelson, D. L., & Simmons, B. L. (2003). Health psychology and work stress: A more positive approach. In J. C. Quick & L. E. Tetrick (Eds.), *Handbook of occupational health psychology* (pp. 97–119).

Nieder, T., & Sieffge-Krenke, I. (2001). Coping with stress in different phases of romantic development. *Journal of Adolescence, 24*(3), 297–311.

Ong, A. D., Bergeman, C. S., Bisconti, T. L., & Wallace, K. A. (2006). Psychological resilience, positive emotions, and successful adaptation to stress in later life. *Journal of Personality and Social Psychology, 91*(4), 730–749.

Powers, S. I., Pietromonaco, P. R., Gunlicks, M., & Sayer, A. (2006). Dating couples' attachment styles and patterns of cortisol reactivity and recovery in response to a relationship conflict. *Journal of Personality and Social Psychology, 90*(4), 613–628.

Pressman, S. D., & Cohen, S. (2005). Does positive affect influence health? *Psychological Bulletin, 131*(6), 925–971.

Reich, J. W., & Zautra, A. (1981). Life events and personal causation: Some relationships with satisfaction and distress. *Journal of Personality and Social Psychology, 41*(5), 1002–1012.

Rietveld, S., & van Beest, I. (2007). Rollercoaster asthma: When positive emotional stress interferes with dyspnea perception. *Behaviour Research and Therapy, 45*(5), 977–987.

Robles, T. F., Shaffer, V. A., Malarkey, W. B., & Kiecolt-Glaser, J. K. (2006). Positive behaviors during marital conflict: Influences on stress hormones. *Journal of Social and Personal Relationships, 23*(2), 305–325.

Sbarra, D. A. (2006). Predicting the onset of emotional recovery following nonmarital relationship dissolution: Survival analyses of sadness and anger. *Personality and Social Psychology Bulletin, 32*(3), 298–312.

Sbarra, D. A., Law, R. W., Lee, L. A., & Mason, A. E. (2009). Marital dissolution and blood pressure reactivity: Evidence for the specificity of emotional intrusion-hyperarousal and task-related emotional difficulty. *Psychosomatic Medicine, 71*(5), 532–540.

Selye, H. (1975). Confusion and controversy in the stress field. *Journal of Human Stress, 1*(2), 37–44.

Selye, H. (1978a). On the real benefits of eustress. *Interviewed by Laurence Cherry, Psychology Today, 12*(March).

Selye, H. (1978b). *The stress of life* (revised ed.). New York: McGraw-Hill.

Snyder, B. K., Roghmann, K. J., & Sigal, L. H. (1993). Stress and psychosocial factors: Effects on primary cellular immune response. *Journal of Behavioral Medicine, 16*(2), 143–161.

Sprecher, S., Felmlee, D., Metts, S., Fehr, B., & Vanni, D. (1998). Factors associated with distress following the breakup of a close relationship. *Journal of Social and Personal Relationships, 15*(6), 791–809.

Stefano, G. B., & Esch, T. (2005). Love and stress. *Neuroendocrinology Letters, 26*(3), 173–174.

Stefano, G. B., Stefano, J. M., & Esch, T. (2008). Anticipatory stress response: A significant commonality in stress, relaxation, pleasure and love responses. *Medical Science Monitor: International Medical Journal of Experimental and Clinical Research, 14*(2), RA17–RA21.

Stone, A. A., Cox, D. S., Valdimarsdottir, H., Jandorf, L., & Neale, J. M. (1987). Evidence that secretory IgA antibody is associated with daily mood. *Journal of Personality and Social Psychology, 52*(5), 988–993.

Stress (biology). (2010, July 3). Retrieved July 5, 2010, from http://en.wikipedia.org/wiki/Stress_(biology)

Tomaka, J., Blascovich, J., Kelsey, R., & Leitten, C. (1993). Subjective, physiological, and behavioral effects of threat and challenge appraisal. *Journal of Personality and Social Psychology, 65*(2), 248–260.

Tomaka, J., Blascovich, J., Kibler, J., & Ernst, J. (1997). Cognitive and physiological antecedents of threat and challenge appraisal. *Journal of Personality and Social Psychology, 73*(1), 63–72.

Tomaka, J., Palacios, R., Schneider, K., Colotla, M., Concha, J., & Herrald, M. (1999). Assertiveness predicts threat and challenge reactions to potential stress among women. *Journal of Personality and Social Psychology, 76*(6), 1008–1021.

Tucker, P., & Aron, A. (1993). Passionate love and marital satisfaction at key transition points in the family life cycle. *Journal of Social & Clinical Psychology, 12*(2), 135–147.

Wheaton, B. (1990). Life transitions, role histories, and mental health. *American Sociological Review, 55*(2), 209–223.

Wise, R. A., & Rompre, P.-P. (1989). Brain dopamine and reward. *Annual Review of Psychology, 40*, 191–225.

Zautra, A. J., & Reich, J. W. (1983). Life events and perceptions of life quality: Developments in a two-factor approach. *Journal of Community Psychology, 11*(2), 121–132.

CHAPTER **10**

Matters of the Heart

Couples' Adjustment to Life Following a Health Crisis[1]

JENNIFER G. LA GUARDIA

University of Rochester

Throughout life, everyone needs close others who will provide comfort and support. In fact, having strong and stable connections to close others and support from close others is so essential to well-being that it is considered a basic psychological need (Baumeister & Leary, 1995; Ryan & Deci, 2000). In addition to close friends, a spouse or romantic partner[2] provides the foundation for connection and security in adulthood. In healthy romantic relationships, partners are accessible and responsive to each other's needs (La Guardia & Patrick, 2008). That is, partners send clear signals about their needs, they tune in to each others' signals, and they respond to each other by providing emotional support and comfort. However, many life stressors may challenge partners' abilities to be open and responsive to each other (Randall & Bodenmann, 2009).

Normatively, everyday stressors (e.g., work, family obligations) impose on partners' time and focus on each other. When the foundation of the

[1] Work by the author cited in this chapter was supported by a University of Waterloo seed grant and a Social Sciences and Humanities Research Council of Canada (SSHRC) Grant #410-2004-0946.

[2] Although not all couples are married, I will use the term *spouse* rather than *partner* throughout this manuscript. This is done because I use the term *partner* or *partners* to refer to processes that apply to both members of the couple, and I use *patient* and *spouse* to distinguish between processes unique to these roles.

relationship is strong, the impact of such stressors can be lessened by partners turning to each other to share the emotional and instrumental load of their stressors. However, if stressors become too intense or persist over time, partners may both fail to signal that they are in need of support and also fail to respond to each other's needs in a timely and appropriate manner. As a consequence, when partners cannot gain access to or soothe each other, life stressors may stretch their bond and may negatively impact their personal and relational health (Johnson, 2004).

Some poignant life events, such as a health crisis of one member of the couple, will significantly challenge even the strongest couples (Waltz, 1986). To use an analogy, emotional interactions between partners can be characterized as a dance. In a dance, each partner has his or her own sense of rhythm (temperament) and set of moves (emotion regulation), and partners take cues from and adjust to each other's moves (dyadic emotion regulation). Before a health crisis, together the couple has a unique tempo to their interactions. Some couples have a well-coordinated tango. Other couples battle for control in a contentious *paso doblé*. Still others have, at best, an awkward waltz. At worst, a couple doesn't interact much at all—in essence, the partners sit out of the dance altogether. When a health crisis occurs, it is as if the music gets cranked up loudly and plays at a frenetic speed (emotional arousal). No matter what their dance looked like before the event, it becomes terribly hard for partners to maintain the energy of the dance and coordinate movements between them. Both partners get quickly exhausted, they end up stepping on each other's feet, and they get pulled in different directions. Eventually they may lose their rhythm altogether. After the immediate crisis is over and the music slows down a bit, the health event doesn't simply go away but instead becomes an intruder that keeps trying to cut in on the couples' dance. At the very least this intruder in the dance makes it exceptionally difficult to maintain a good connection and coordinate moves together. Keeping this intruder (the health event) in the dance may worsen any negative patterns of coping that existed in the couple before the health crisis, or it might even create new negative cycles that become problematic for the couple. As such, a health event can impact partners' ability to connect in a time when they are most in need of each other.

Health events therefore have the potential to increase the frequency and intensity of many negative emotions (e.g., fear, worry, sadness) for *both* partners, and with increased distress partners may have greater difficulties in managing their own emotions and the couple may have greater difficulty connecting to draw on their collective strength to cope with the health event (Johnson, 2002). Additionally, for those couples who have difficulty being responsive to each other prior to the health event, the challenges that follow from a health crisis event can be even more difficult to navigate.

Understanding couples' interactions after a health event is critical to understanding the subsequent health of the partners and the health of the relationship. That is, in the aftermath of a health event, partners can influence each other's psychological health (e.g., Badr, Acitelli, & Carmack Taylor, 2008; Fang, Manne, & Pape, 2001; Manne, Ostroff, Norton, Fox, Goldstein, & Grana, 2006; Newsom & Schulz, 1998) and physical health (Dorros, Card, Segrin, & Badger, 2010; Robles & Kiecolt-Glaser, 2003). By better understanding couples' emotional dynamics we may begin to identify potential avenues for intervention to change these interactions before negative psychological and physical health consequences emerge.

In this chapter I will review the ways both partners can become significantly challenged when one of the members of the couple becomes ill. Specifically, I will review how health events increase the frequency and intensity of many negative emotions (e.g., fear, worry, sadness) for both patients and spouses and how this increased distress may make it difficult for each member of the couple to effectively manage their own emotions let alone respond with support to the other. I focus up front on the challenges of acute cardiovascular crises as an illustrative example and then draw on the substantial work from the cancer and congestive heart failure literatures to examine how couples' coping dynamics impact functioning in both partners and thereby impact the climate of recovery.[3] Finally, I argue for a couples-based, emotion-focused approach to intervention for recovery after health crises and discuss potential positive repercussions for such an approach on psychological and physical health of *both* partners. Let me first begin by providing a bit of background on some of the challenges that couples face in the aftermath of an acute cardiovascular crisis to better illustrate the systemic emotional upheaval that occurs for couples after such an event.

Impact of Acute Health Events on the Couple

Impact of a Cardiac Event on the Patient

To appreciate why patients might have significant emotional distress both in an acute cardiac crisis and afterward in recovery, let's take a moment

[3] While there are nontrivial differences between acute health events that have chronic health care consequences (e.g., heart attack, cardiac bypass surgery) and those health conditions for which treatment unfolds over long periods of time (e.g., breast cancer, prostate cancer) or that are deteriorative with poorer prognosis (e.g., congestive heart failure, lung cancer), there are many more similarities in how these events shift both the emotional set within individuals and shift the relational dynamics of the couple. That is, as a result of any of these health events, there are often significant changes in patients' everyday functioning, partners' role responsibilities, the couples' lifestyle behaviors, and the couples' intimacy, and both patients and spouses have strong emotional reactions that they have to actively cope with.

to walk in their shoes. For those who have had a heart attack, the experience ranges from feeling numbness in their arm, an inability to catch their breath, or extreme fatigue and nausea, to experiencing crushing pressure and chest pain (described often as an elephant standing on one's chest), being rushed to the emergency room, and having to be resuscitated by CPR. After a heart attack, patients often immediately undergo a revascularization surgery (stent, angioplasty, coronary artery bypass graft surgery), which carries its own associated survival risks. Patients are understandably frightened and worried.

When patients come out of surgery, postsurgical intensive care and recovery is a parade of poking, prodding, and intense monitoring. Then, just as soon as the patient can re-establish very basic functioning (typically within only a few days of surgery), they are sent home with their family, without the monitors, nurses, or doctors that they have come to rely on for information and reassurance about their health. As such, a whole new cycle of fear and worry is born.

Given these experiences, it is no wonder that patients normatively have significant psychological distress during the acute crisis and may have residual distress afterward (Al-Hassan & Sagr, 2002; Condon & McCarthy, 2006). Indeed, after a cardiac event, patients normally feel significant fears about death, and they worry about their future health (e.g., Will I have another cardiac event? Will I ever be healthy and able to function fully again?). These worries cause patients to be quite vigilant about their bodily sensations, so that a tingling in their arm, a faster heartbeat, or difficulty catching their breath might send them into a panic that their heart will give out again. Patients also often feel overwhelmed by lifestyle changes such as altering their diet and exercise, taking medications, and quitting smoking. Any one of these alone would be difficult, and for some patients, these are quite drastic changes from their previous lifestyle. Patients may also be frustrated by physical limitations, as it often takes much more time and energy for them to engage in many routine activities (e.g., walking, climbing stairs, going to the grocery store, cooking a meal). For those who have to return to work, the pace and the types of activities that they are responsible for may be reduced for some time, if not permanently. If the patient has to slow down and cannot take care of the things he or she normally would do, role shifts also have to occur. Family members, particularly the spouse, take up the duties that would normally be the patients' or be shared by the couple. Although for some patients this role shift can be a welcomed break from responsibility it is more often met with mixed feelings. For many it is the first time in a long time that they are dependent on others to take care of things that they have been normally able to do on their own. Further, for many, the roles they inhabit are an important part of their identity, allow them to feel competent, and allow them to "contribute" or show their care for others. Thus, altering their roles or giving them up altogether can feel like a huge loss and add another significant layer of stress to their recovery.

For a significant number of patients, a combination of these stressors contributes to the development of psychopathology. For example, for those who have had a heart attack, approximately 25 to 37% develop clinically significant depression (Kaptein, De Jonge, van den Brink, & Korf, 2006; Lane, Carroll, Ring, Beevers, & Lip, 2002), 40% develop anxiety (Lane et al.), and 10 to 22% develop posttraumatic stress symptoms (Bennett, Conway, Clatworthy, Brooke, & Owen, 2001; Ginzburg et al., 2003; Shemesh et al., 2001). Importantly, having persistent depression, anxiety, and posttraumatic stress symptoms is significant for patients' physical recovery and future health. For example, those who continue to have depressive symptoms during recovery are at greater risk for having new cardiovascular events (e.g., another heart attack; Kaptein et al., 2006), more infections (e.g., pneumonias, upper respiratory infections) and other illnesses (Doering, Martínez-Maza, Vredevoe, & Cowan, 2008), and for dying early (Barth, Schumacher, & Herrmann-Lingen, 2004). Patients that have significant persistent anxiety are also at greater risk for having new cardiovascular events and are more likely to seek medical care for health problems (Strik, Denollet, Lousberg, & Honig, 2003). Finally, those with higher rates of posttraumatic symptoms are less likely to take control of their health behaviors (e.g., taking medications for managing their blood pressure and cholesterol; quitting smoking; Shemesh et al., 2001, 2004). In summary, patients have many stressors that emerge after their cardiac events, and, for many, the persistence of these stressors will have a significant impact on their psychological and physical health.

Impact of a Cardiac Event on the Spouse

One of the key relationships that patients rely on after a cardiac crisis is the one with their spouse. However, spouses often have their own challenges in the recovery period that may bring on significant distress and limit their ability to be supportive of patients. In the acute crisis, spouses have their own intense emotions about the cardiac event, as they are often witness to the patient's actual cardiac event and have to endure uncertainty about whether the patient will live or die. Once the acute crisis passes, fears about the patient dying still remain, and additional worries about the patient's treatment, recovery, and long-term prognosis become extraordinarily salient. Upon the patient's discharge from the hospital, the spouse typically takes up the role of primary caregiver, providing both emotional and instrumental support to the patient (e.g., cooking, cleaning, tracking the patient's medications). As previously mentioned, spouses often take up family responsibilities that may have once been shared. If these stressors were not enough, spouses also are asked to make significant lifestyle changes (e.g., modify their diet, exercise, quit smoking) to accommodate patients' new health regimens. Just like patients, most spouses will have mixed feelings about these changes, as they are often willing to make the changes but don't initially like to or want to change.

Spouses' distress has been shown to equal or surpass the distress felt by patients across the first year after hospitalization (Artinian, 1991, 1992; Moser & Dracup, 2004; O'Farrell, Murray, & Hotz, 2000). Spouses who are distressed are at higher risk for developing depression and anxiety (Beach, Schulz, Yee, & Jackson, 2000; O'Farrell et al.) as well as for developing greater physical symptoms and engaging in more health risk behaviors (Beach et al.). In fact, among elderly spouses the emotional strain of caregiving is an independent risk factor for early mortality (Schulz & Beach, 1999). Thus, the cardiac event can clearly cause significant emotional distress in spouses and can significantly impact spouses' psychological and physical health.

Impact of a Cardiac Event on the Partnership

As the cardiac event takes center stage in the couple's life, it creates added stress on the partnership as well (Mahrer-Imhoff, Hoffmann, & Froelicher, 2007; O'Farrell et al., 2000). Although those who are in stronger relationships before a cardiac event tend to function better after a cardiac event than those who are in distressed marriages before such an event (Brecht, Dracup, Moser, & Riegel, 1994; Waltz, Badura, Pfaff, & Schott, 1988), approximately one-third of couples who are functioning well prior to a cardiac event show declines in their relationship functioning over the first year after the event (Waltz, 1986).

The many challenges that have already been discussed impact not only the individual partners but also the partnership. Beyond these challenges, cardiac events can further intrude on the couple's physical intimacy by amplifying difficulties couples were already having around being physically connected or by bringing out new fears and vulnerabilities that may not have been present prior to the event (O'Farrell et al., 2000). Indeed, sexual problems are common after heart attack and bypass surgery, occurring in approximately one-half to three-quarters of all patients (Birkhauser, 2009; Kloner et al., 2003). Although some of this is a result of physical restrictions of the patient after the cardiac event, some can also be attributed to the interactions of the couple. For example, after a cardiac event, patients and their partners often have significant fear that sexual contact will harm the patient (Beach et al., 1992; Lindau et al., 2010). These fears can get in the way of partners initiating and engaging each other sexually, and this reduced contact can also uniquely contribute to loss of interest or desire as well as difficulties getting sexually aroused in one or both partners. Thus, when both partners may want and need to get closer to each other, they may refrain from doing so.

The importance of a strong, stable, and supportive relationship for recovery and long-term health cannot be understated. Across the general population as well as those specifically coping with illness, marital quality is a significant predictor of better health outcomes (Robles & Kiecolt-Glaser, 2003). Specifically with regard to cardiac events, marital quality has been shown to be a potent predictor of psychological health after a cardiac event (e.g., Brecht et al., 1994; Waltz et al., 1988), length of hospital stay after coronary artery bypass graft (CABG) surgery (e.g., Kulik & Mahler, 2006), recurrence of cardiac events after a heart attack (e.g., Orth-Gomer, Wamala, Horsten, Schenck-Gustafsson, Schneiderman, & Mittleman, 2000), and rate of survival (King & Reis, 2011; Rohrbaugh, Shoham, & Coyne, 2006). Thus, the couple's relationship can be an important buffer after a health event, or it may contribute to greater difficulties in recovery.

Why Is Attending to Emotions So Important After a Health Event?

From the example of cardiac events it should be clear that couples face significant emotional challenges in the aftermath of a health event. How partners cope with their emotions about a health event thus has important implications for their personal and relational functioning. Partners can attempt to reduce overwhelming negative emotions about the health event by intrapersonally or interpersonally regulating their emotions. For example, partners might try to reduce their negative affect by trying to ignore or distract themselves from the emotions that are emerging or by trying to actively suppress their negative emotional experiences (*intrapersonal* level). Moreover, even when they allow themselves to experience their negative emotions, as a way to prevent negative emotions from escalating they may attempt to limit how much they disclose their negative feelings about the health event to their partner (*interpersonal* level).

Although these regulatory strategies may be employed in an effort to reduce feelings of distress, research has shown that limiting one's emotional experience and expression actually has significant personal costs (e.g., Butler, Egloff, Wilhelm, Smith, Erickson, & Gross, 2003; Gross & John, 2003; Mauss & Gross, 2004; Pennebaker, 1997). For example, cardiac patients with a trait tendency to experience greater negative affect (e.g., be chronically dysphoric, worried, irritable) and be socially inhibited (e.g., withdrawn, less expressive of thoughts and feelings, limited in social interactions to avoid disapproval), called Type D personality (Denollet, 2000), show lower quality of life (Denollet, Vaes, & Brutsaert, 2000) and are at greater risk for subsequent adverse cardiac events and early mortality (e.g., Denollet et al., 2006). Beyond dispositional coping strategies, how patients and their partners cope with their feelings specifically *about a health event* when with each other has been shown to have important implications for psychological health as well. For example, after a heart attack, patients who hide or deny worries about their heart attack have higher distress over 6 months post-hospitalization (Suls, Green, Rose, Lounsbury, & Gordon, 1997) and those who are less disclosing to their spouse about their feelings about their heart attack experience more severe chest pain, are more likely to be rehospitalized, and report decreased health 1 year after hospitalization (Helgeson, 1991). Further, in our recent survey of patients who experienced a heart attack, coronary artery bypass graft surgery, or both within the past year, the more patients tried to suppress, ignore, or distract themselves from their feelings about their cardiac event when with their spouse, the more they experienced symptoms of depression, anxiety, and posttraumatic stress (So & La Guardia, 2011). Additionally, the more spouses tried to inhibit their own negative feelings about the cardiac event, the more psychological distress they also felt (La Guardia & So, 2010). Evidence from the cancer literature also strongly supports these findings, such that the more patients and their spouses make cancer-related emotional disclosures, the better their psychological adjustment (see Badr et al., 2008). In summary, both patients and spouses are psychologically healthier the more that they are effectively coping with their own feelings that come up about a health event.

Partners are interdependent and influence each other in meaningful ways (Kelley & Thibault, 1978; Rusbult & Arriaga, 2000). As such, effectively coping with one's own emotions may impact not just one's own health but also the health of one's partner and the functioning of the relationship. When partners cope effectively, they are better able to call on each other for support, and they are better able to be available and responsive to each other when in need (Gottman, 1994; Johnson, 2004). In contrast, when partners are occupied with trying to reduce their own distress, partners may behave in ways that actually undermine their connection to each other and thus undercut their ability to draw on the collective coping strength of their partnership (Gottman, 1994; Johnson, 2004). To understand these behavioral dynamics, let's turn to a discussion of the behaviors partners might engage in when a health event makes the threat to the patient's life and thus the threat to the relationship more salient.

Emotional Responsiveness in Good Relationships

When a cardiac event occurs, there is a real danger that the patient will die and the couple will thus actually lose their relationship. The potential of this disconnection often triggers a strong emotional response, flooding partners with emotions, especially negative emotions such as fear. The emotion system is homeostatic by nature and thus responds to this arousal by mobilizing resources to reduce the distress. Partners may attempt to reduce their distress in many different ways. They may respond to their distress by getting closer to each other and drawing on their connection to cope or they may attempt to quiet their distress in ways that actually get in the way of their connection and ironically undermine their ability to cope. Herein I outline reactions that pertain to both partners as well as a few reactions that are unique to spouses (and other non-ill partners).[4]

Partners may respond to their distress and attempt to reduce their fear of losing the relationship by *seeking proximity* to each other (Mikulincer, Florian, & Hirschberger, 2004). That is, partners will reach out to each other and try to get close to restore a sense of connection and safety. They may do this by directly expressing affection through their words ("I love you") and their physical touch (e.g., kisses, hugs, holding hands), by spending more time with each other, and by simply making their relationship a priority. By drawing closer to each other, partners are available to be more emotionally and instrumentally supportive of each other when stressors arise. Indeed, after a cardiac event many partners will reconnect and become more attuned to each others' needs, and this renewed connection helps couples to collectively take on the challenges of cardiac disease (Mahrer-Imhof et al., 2007). Further, after a cancer diagnosis, the more that spouses are actively engaged with patients (e.g., asking what the patient feels and needs), the more satisfied patients are with their marriage and the better they are able to cope with treatment (Hagedoorn et al., 2000). Importantly, this benefit is most pronounced for patients who are the most highly distressed and thus are most in need of support.

Partners may also react to their distress and attempt to quiet their fears in ways that actually disrupt their connection and distance them from each other just when they need each other the most (Rafaeli & Gleason, 2009). These behaviors include *withdrawing from or avoiding one's partner* and *getting irritated or hostile with one's partner* (Gottman, 1994; Johnson, 2004). First, sometimes being with a partner can make one more aware of fears or vulnerabilities, so as a way to make these fears or vulnerabilities feel less overwhelming one may try to escape these feelings or avoid talking about them by withdrawing from conversations with one's partner or by avoiding contact with one's partner altogether. Although this strategy may serve to temporarily reduce one's distress, not only does it make one less available to receive support by one's

[4] Notably, the vast majority of the literature examines how spouses' reactions to their own distress impact patients' functioning and there is very little on how patients' emotional reactions to their own illness impact spouses' functioning. Thus, the examples reviewed here focus on spouses' reactions to their own distress and the impact of their behaviors on patients. Clearly, patients' reactions to their own distress and the impact of their own behaviors on spouses' functioning is a critical area for future study.

partner, but evidence also suggests that it has a negative impact on one's partner. Indeed, when spouses are avoidant or convey discomfort when patients try to talk about their illness (a passive strategy to shut down the conversation), cancer patients become more distressed (Manne, Ostroff, Winkel, Grana, & Fox, 2005). Further, in cardiac patients, when spouses are less available to be supportive, patients are less compliant with recommended health behaviors, they show slower physical recovery, they are more psychologically distressed, and they have lower overall well-being (Beach et al., 1992; Coyne & Smith, 1991; Moore, 1994; Moser & Dracup, 2004; Thompson & Cordle, 1988).

Partners may also react to their distress and create more distance in their relationship when feelings of fear or vulnerability become intense or when negative feelings have built up, and they start to leak out or boil over in a fury. Oftentimes, a partner will get the brunt of this distress, such that one will get easily irritated with or critical of him or her. Again, not only does reacting in this way push one's partner away and make one less likely to receive support, evidence shows that this behavior also has a negative impact on one's partner. For example, the more wives were hostile toward their husbands after the husbands' cardiac event, the more the husbands tended to be psychologically distressed (Fiske, Coyne, & Smith, 1991). Further, when spouses were more hostile or criticized cancer patients for how they were coping with their illness, patients were more distressed and found engaging in the relationship less satisfying (Manne, Ostroff, Winkel, Grana, & Fox, 2005). Importantly, whether spouses engage in either avoidant or hostile behaviors, evidence suggests that patients become distressed by their spouse's unsupportive behaviors in part because these behaviors lead patients to avoid coping effectively with their own stress about their illness (e.g., they become less disclosing; Manne, Ostroff, Winkel, et al., 2005). Thus, spouses' coping impacts patients' coping in recovery.

Spouses may also have some other unique responses when the patients' health is threatened. First, spouses may become more *hypervigilant* about the patients' health (Theobald & McMurray, 2004). That is, a spouse might actively worry and check in on the patient by monitoring the patient or by frequently asking the patient if he or she is okay. Another way that spouses might respond to their own fears is by trying to directly limit patients' behaviors or take over tasks for them, even when patients are capable (e.g., Franks, Stephens, Rook, Franklin, Keteyian, & Artinian, 2006). That is, spouses might try to restructure and direct the patient's daily activities, shield the patient from life stressors (e.g., protective buffering), or actively stop the patient from engaging in everyday tasks in order to exert some control and attempt to prevent further harm to the patient (e.g., if I lessen my partner's exertion or stress, it will prevent him from having another cardiac event or dying). While some evidence suggests that spouses' being hypervigilant and taking over for patients can serve an adaptive function in recovery (Fiske et al., 1991; Gilliss, 1984), the vast majority of evidence suggests that these behaviors, although well intentioned, may undermine patients' health and recovery (e.g., Clarke, Walker, & Cuddy, 1996; Coyne & Smith, 1991; Franks et al., 2006; Goldsmith, Lindholm, & Bute, 2006; Newsom & Schulz, 1998; Tucker, 2002). For example, in cancer patients, the more spouses are overprotective (Hagedoorn et al., 2000) or try to solve problems *for* patients rather than *with* them (Manne et al., 2004), the more distressed patients are. Further, in congestive heart failure patients, the more spouses are overinvolved, the more distressed and the less efficacious patients feel in their daily lives (Benazon, Foster, & Coyne, 2006). Finally, the more spouses are controlling of patients' health-related behaviors (e.g., diet, exercise) the less patients engage in these healthy behaviors and the poorer patients' psychological health over time (Franks et al.).

Whether spouses' behaviors are supportive or whether they undermine patients may depend on whether spouses' behaviors are actually in response to the patients' needs or whether spouses' behaviors are being driven by their own need to down-regulate their own negative emotions. For example, in our own recent work with cardiac patients and their spouses, we found that the more spouses are distressed and not coping well with their own feelings about a cardiac event, they more likely they are to be hypervigilant of the patients' health and controlling of the patients' behaviors (La Guardia & So, 2011). Importantly, spouses' behaviors were not simply about patients' need for support (as indexed by patients' levels of anxiety, depression, and posttraumatic stress but rather were a reaction to their difficulties in dealing with their own negative emotions about the cardiac event. Thus, hypervigilance and control can be exacerbated when spouses are not regulating their own emotions well, thereby undermining the climate of recovery for the patient.

In summary, when there is a threat of physical harm to a partner, as in a health crisis, partners will experience significant distress and will engage in various behaviors to reduce this distress. Seeking proximity to a partner can serve to buffer stress and engage the couples' collective strength for coping. Unfortunately, however, sometimes when partners struggle to deal with painful feelings, they undermine their relationship connection just when they need it the most. It is clear that patients' health is intimately tied to spouses' emotional coping after the crisis. What remains to be better understood is how patients' emotional coping impacts spouses.

Importance of the Relational Climate for Both Partners After a Health Crisis

As I mentioned, although the vast majority of research has focused on how one partner's behavior affects the other (particularly how spouses' behaviors impact patients), recent research has begun to look more concertedly at how the coping *dynamics* of the couple impact partners' functioning after illness has struck. How relational partners collectively cope with illness has been captured in the construct of "relationship talk." Relationship talk reflects the extent to which couples discuss the state of their relationship, each partner's needs, and the relational implications of a shared stressor (Acitelli & Badr, 2005). Overall in relationships, relationship talk is thought to help couples define their relationship and repair their relationship when it is poorly functioning. It has been shown to be particularly important for adjustment in couples coping with chronic illness. For example, when discussing a wife's breast cancer, reciprocal self-disclosure between the patient and her spouse is associated with lower general and cancer-specific distress as well as lower incidence of cancer-related intrusions and avoidance in the patient (Manne et al., 2004). Also, for patients with congestive heart failure, a communal orientation (holding a shared assessment of a health threat and collective action in managing the threat) as represented by "we" talk of partners when discussing the illness has positive implications for the health of the patient (Rohrbaugh, Mehl, Shoham, Reilly, & Ewy, 2008). In couples in which one partner has been diagnosed with lung cancer, the more the couple talks about the relationship implications of the lung cancer, the lower the distress in both partners and the greater their marital satisfaction 6 months after initial diagnosis and treatment (Badr et al., 2008). Even more interesting, this benefit was most pronounced for the non-ill partner. Finally, in couples coping with cancer, greater mutual constructive communication about cancer-related problems is associated with less distress and greater marital satisfaction for both partners (Manne, Ostroff, Norton, et al., 2006). In contrast, attempts to avoid discussing personal concerns or the cancer itself have been shown to be associated with greater distress in both partners (e.g., Manne, Norton, et al., 2007).

In summary, it is clear that health events are *systemic* events. That is, health events emotionally impact both patients and spouses, and optimal health in recovering from such events requires both partners to be available and engaged. Given this knowledge, treatment interventions presumably should focus on ways to help partners join together to cope with the many challenges that they face in the aftermath of a health crisis. Unfortunately, to date, in most cases the patient is the sole focus of intervention.

Couples Interventions

Currently, many interventions following health events focus almost exclusively on improving patients' *physical* health, and those interventions that have addressed *psychological* health typically employ short-term individual psychotherapy or stress management programs for the patient only (e.g., see Clark, Hartling, Vandermeer, & McAlister, 2005; Lepore & Coyne, 2006; Linden, Phillips, & Leclerc, 2007 for reviews). To date, these psychosocial interventions have shown limited success in reducing morbidity and mortality, with some notable exceptions (e.g., Classen et al., 2001; Denollet & Brutsaert, 2001; Manne, Rubin, et al., 2007; Rollman et al., 2009).

Couples-based psychosocial interventions clearly shift current clinical practice paradigms toward a systemic approach to recovery and offer several advantages over existing patient-focused interventions. First, as the evidence reviewed earlier suggests, there are significant psychosocial consequences of health events for both patients and spouses. Thus, interventions that focus on the couple may benefit both patients and spouses psychologically, and by improving their communication couples will have the tools to better collaboratively tackle instrumental challenges (e.g., health behavior changes). Thus, couples interventions would be in line with the aims of existing psychosocial and behavior change interventions for patients but would also afford additional support to spouses and to the relationship.

Second, couples interventions potentially have important implications for the rehabilitation of women patients. For example, for cardiovascular disease, women carry a larger burden of illness, regardless of whether they are the patient or the caregiver (Coyne & Fiske, 1992; Lyons, Sullivan, Ritvo, & Coyne, 1995). Although women have a high rate of coronary heart disease (CHD; with a 32% lifetime risk of developing CHD after age 40), they are 55% less likely than men to be referred to and participate in cardiac rehabilitation (AHA, 2009), and they are more likely to feel compelled to more quickly resume their roles and responsibilities after a cardiac event, even to the neglect of their own recovery needs (Kristofferzon, Lofmark, & Carlsson, 2003). Given that women's quality of life is much worse than men's following cardiac events (Emery et al., 2004) and some evidence suggesting that marital quality is key to predicting women's survival after a cardiac event (Rohrbaugh et al., 2006), employing a couples intervention may be particularly critical for women's health.

Couples-based interventions that at least in part focus on partners' emotional communication have been developed for cardiac couples and show some promise (e.g., Sher & Baucom, 2001). However, much of this type of couples' intervention work has been conducted in the treatment of women's cancer (see Manne & Ostroff, 2008; Scott, Halford, & Ward, 2004). For example, for couples in which the woman has early-stage breast cancer, Manne, Ostroff, Winkel, Fox, et al. (2005) demonstrated the benefit of using a couples-focused group intervention over typical psychosocial care. Specifically, the intervention—aimed at enhancing support exchanges and coping skills in couples—was shown to reduce depression and anxiety, to increase behavioral and emotional control, and to increase well-being in patients, particularly for those women who rated their partners as more unsupportive pre-intervention or for those who reported greater physical impairment at baseline. In another example, for couples in which the woman has early-stage cancer (breast or gynecological), Scott et al. (2004) demonstrated the benefit of a couples' coping training intervention (CanCOPE) over medical information education training and patient-only coping training. The intervention—aimed at improving couples communication about cancer (learning how to more effectively self-disclose and empathically listen and validate each other)—increased observable supportive communications in couples and was also associated with both partners reporting reduced effort in their coping (in essence, they better "shared the affective load"). Further, for patients, the intervention decreased their psychological distress, reduced their need to avoid intrusive negative thoughts about cancer, improved their self-perceptions about their sexuality, and fostered greater intimacy in their relationship.

In summary, some evidence is beginning to amass to suggest that intervening with the couple rather than just the patient has important implications for optimizing recovery. The goal of future work should be to further develop couples-based interventions and to adapt those interventions that show promise to treat other acute and chronic illness.

Conclusions and Future Directions

Health crises are frightening for both patients and spouses, and the many stressors that emerge from a health event can be extraordinarily difficult for couples to navigate effectively. Indeed, health crises can challenge even those partners who were good at managing their feelings and being supportive of each other before the health event. Given that couples face significant emotional challenges in the aftermath of a cardiac crisis, how they collectively cope with their emotions has important implications for their health and relational functioning.

Couples-based interventions provide an opportunity to promote healthier coping in patients and their spouses and thereby prevent some of the potential psychological and physical costs of such health events. In designing and improving these interventions it will be important to draw on existing couples interventions from the clinical literature—ones that have sound theoretical foundations and solid empirical evidence for their effectiveness with couples with diverse challenges (e.g., emotion-focused therapy; Johnson, 2004). Such interventions are essentially attuned to the emotional dynamics of the couple and thus provide an important framework for intervening after an emotionally evocative event such as health crisis.

There are several other important considerations for future research. First, given that partners' behaviors are interdependent and that much of the work in the health literature focuses on how spouses' behaviors impact patients, greater attention should be paid in future work to how both partners' coping affects their own and each other's psychological and physical health outcomes. This requires that in our research designs we attend to *assessing* coping processes in both members of the couple and *modeling* both members' outcomes as a function of their own and their partner's behaviors (e.g., using actor-partner modeling or structural equation modeling modeling for distinguishable partner dyads).

Second, as mentioned previously, gender is an important variable to consider in modeling the behaviors of couples. Women have been shown to behave somewhat differently than men, whether in the role of patient or in the role of the caregiver (e.g., see Revenson, Abraido-Lanza, Majerovitz, & Jordan, 2005). Although I predict that the emotion regulation dynamics described in this paper have similar functional implications for health for women and men, future research will absolutely need to consider whether gender differences are evident.

Finally, as is common in any field, we often frame our research questions as unidirectional rather than look at the bidirectional influence of how disease shapes psychosocial functioning and how psychosocial functioning also shapes the progression of disease. This biopsychosocial perspective (Engel, 1977) becomes particularly crucial after a health event, as much evidence suggests that there is a bidirectional influence between biological functioning (e.g., neurophysiological, immunological, endocrine) and psychosocial functioning. Thus, future research will benefit from an interdisclipinary eye to research design and measurement to account for this interplay.

References

Acitelli, L. K., & Badr, H. J. (2005). My illness or our illness? Attending to the relationship when one partner is ill. In T. A. Revenson, K. Kayser, & G. Bodenmann (Eds.), *Couples coping with stress: Emerging perspectives on dyadic coping* (pp. 137–156). Washington, DC: American Psychological Association.

Al-Hassan, M., & Sagr, L. (2002). Stress and stressors of myocardial infarction patients in the early period after discharge. *Journal of Advanced Nursing, 40*(2), 181–188.

American Heart Association (AHA). (2009). *Heart disease and stroke statistics update.* Dallas, TX: Author.

Artinian, N. T. (1991). The stress experience of spouses of patients having coronary artery bypass during hospitalization and six weeks after discharge. *Heart & Lung, 20*(1), 52–59.

Artinian, N. T. (1992). Spouse adaptation to mate's CABG surgery: 1-year follow-up. *American Journal of Critical Care, 1*, 36–42.

Badr, H., Acitelli, L. K., & Carmack Taylor, C. L. (2008). Does talking about their relationship affect couples' marital and psychological adjustment to lung cancer? *Journal of Cancer Survivorship, 2,* 53–64.

Barth, J., Schumacher, M., & Herrmann-Lingen, C. (2004). Depression as a risk factor for mortality in patients with coronary heart disease: A meta-analysis. *Psychosomatic Medicine, 66,* 802–813.

Baumeister, R. F., & Leary, M. R. (1995). The need to belong: Desire for interpersonal attachments as a fundamental human motivation. *Psychological Bulletin, 117,* 497–529.

Beach, E. K., Maloney, B. H., Plocica, A. R., Sherry, S. E., Weaver, M., Luthringer, L., et al. (1992). The spouse: A factor in recovery after acute myocardial infarction. *Heart & Lung, 21*(1), 30–38.

Beach, S. R., Schulz, R., Yee, J. L., & Jackson, S. (2000). Negative and positive health effects of caring for a disabled spouse: Longitudinal findings from the Caregiver Health Effects study. *Psychology & Aging, 15,* 259–271.

Benazon, N. R., Foster, M. D., & Coyne, J. C. (2006). Expressed emotion, adaptation and patient survival among couples coping with chronic heart failure. *Journal of Family Psychology, 20,* 328–334.

Bennett, P., Conway, M., Clatworthy, J., Brooke, S., & Owen, R. (2001). Predicting post-traumatic symptoms in cardiac patients. *Heart & Lung: Journal of Acute and Critical Care, 30,* 458–465.

Birkhauser, M. H. (2009). Quality of life and sexuality issues in aging women. *Climacteric, 12*(1), 52–57.

Brecht, M., Dracup, K., Moser, D. K., & Riegel, B. (1994). The relationship of marital quality and psychosocial adjustment to heart disease. *Journal of Cardiovascular Nursing, 9*(1), 74–85.

Butler, E. A., Egloff, B., Wilhelm, F. H., Smith, N. C., Erickson, E. A., & Gross, J. J. (2003). The social consequences of expressive suppression. *Emotion, 3,* 48–67.

Clark, A. M., Hartling, L., Vandermeer, B., & McAlister, F. A. (2005). Meta-analysis: Secondary prevention programs for patients with coronary artery disease. *Annals of Internal Medicine, 143,* 659–672.

Clarke, D. E., Walker, J. R., & Cuddy, T. E. (1996). The role of perceived overprotectiveness in recovery 3 months after myocardial infarction. *Journal of Cardiopulmonary Rehabilitation, 16*(6), 372–377.

Classen, C., Butler, L. D., Koopman, C., Miller, E., DiMiceli, D., Giese-Davis, J., et al. (2001). Supportive-expressive group therapy and distress in patients with metastatic breast cancer: A randomized clinical intervention trial. *Archives of General Psychiatry, 58,* 494–501.

Condon, C., & McCarthy, G. (2006). Lifestyle changes following acute myocardial infarction: Patients perspectives. *European Journal of Cardiovascular Nursing, 5,* 37–44.

Coyne, J. C., & Fiske, V. (1992). Couples coping with chronic and catastrophic illness. In T. J. Akamatsu, M. A. P. Stephens, S. E. Hobfoll, & J. H. Crowther (Eds.), *Family health psychology* (pp. 129–149). Washington, DC: Hemisphere.

Coyne, J. C., & Smith, D. A. F. (1991). Couples coping with a myocardial infarction: A contextual perspective on wives' distress. *Journal of Personality & Social Psychology, 61*(3), 404–412.

Denollet, J. (2000). Type D personality: A potential risk factor refined. *Journal of Psychosomatic Research, 49,* 255–266.

Denollet, J., & Brutsaert, D. L. (2001). Reducing emotional distress improves prognosis in coronary heart disease: 9-year mortality in a clinical trial of rehabilitation. *Circulation, 104*(17), 2018–2023.

Denollet, J., Pedersen, S. S., Ong, A. T. L., Erdman, R. A. M., Serruys, P. W., & van Domburg, R. T. (2006). Social inhibition modulates the effect of negative emotions on cardiac prognosis following percutaneous coronary intervention in the drug-eluting stent era. *European Heart Journal, 27*(2), 171–177.

Denollet, J., Vaes, J., & Brutsaert, D. L. (2000). Inadequate response to treatment in coronary heart disease: Adverse effects of Type D personality and younger age on 5-year prognosis and quality of life. *Circulation, 102,* 630–635.

Doering, L. V., Martínez-Maza, O., Vredevoe, D. L., & Cowan, M. J. (2008). Relation of depression, natural killer cell function, and infections after coronary artery bypass in women. *European Journal of Cardiovascular Nursing, 7*(1), 52–58.

Dorros, S. M., Card, N. A., Segrin, C., & Badger, T. A. (2010). Interdependence in women with breast cancer and their partners: An interindividual model of distress. *Journal of Consulting and Clinical Psychology, 78*(1), 121–125.

Emery, C. F., Frid, D. J., Engebretson, T. O., Alonzo, A. A., Fish, A., Ferketich, A. K., et al. (2004). Gender differences in quality of life among cardiac patients. *Psychosomatic Medicine, 66,* 190–197.

Engel, G. L. (1977). The need for a new medical model: A challenge for biomedicine. *Science, 196,* 129–136.

Fang, C. Y., Manne, S. L., & Pape, S. J. (2001). Functional impairment, marital quality, and patient psychological distress as predictors of psychological distress among cancer patients' spouses. *Health Psychology, 20,* 452–457.

Fiske, V., Coyne, J. C., & Smith, D. A. (1991). Couples coping with myocardial infarction: An empirical reconsideration of the role of overprotectiveness. *Journal of Family Psychology, 5,* 4–20.

Franks, M. M., Stephens, M. A. P., Rook, K. S., Franklin, B. A., Keteyian, S. J., & Artinian, N. T. (2006). Spouses' provision of health-related support and control to patients participating in cardiac rehabilitation. *Journal of Family Psychology, 20*(2), 311–318.

Gilliss, C. L. (1984). Reducing family stress during and after coronary artery bypass surgery. *Nursing Clinics of North America, 19,* 103–111.

Ginzburg, K., Solomon, Z., Koifman, B., Keren, G., Roth, A., Kriwisky, M., et al. (2003). Trajectories of posttraumatic stress disorder following myocardial infarction: A prospective study. *Journal of Clinical Psychiatry, 64*(10), 1217–1223.

Goldsmith, D. J., Lindholm, K. A., & Bute, J. J. (2006). Dilemmas of talking about lifestyle changes among couples coping with a cardiac event. *Social Science & Medicine, 8,* 2079–2090.

Gottman, J. (1994). *What predicts divorce?* Hillsdale, NJ: Lawrence Erlbaum Associates.

Gross, J. J., & John, O. P. (2003). Individual differences in two emotion regulation processes: Implications for affect, relationships, and well-being. *Journal of Personality and Social Psychology, 85*, 348–362.

Hagedoorn, M., Kuijer, R. G., Buunk, B. P., DeJong, G. M., Wobbes, T., & Sanderman, R. (2000). Marital satisfaction in patients with cancer: Does support from intimate partners benefit those who need it the most? *Health Psychology, 19*, 274–282.

Hagedoorn, M., Sanderman, R., Bolks, H. N., Tuinstra, J., & Coyne, J. C. (2008). Distress in couples coping with cancer: A meta-analysis and critical review of role and gender effects. *Psychological Bulletin*, 134, 1–30.

Hall, K. (2004). *Reclaiming your sexual self: How to bring desire back into your life.* New York: John Wiley and Sons.

Helgeson, V. S. (1991). The effects of masculinity and social support on recovery from myocardial infarction. *Psychosomatic Medicine, 53*, 621–633.

Johnson, S. M. (2002). *Emotionally focused couple therapy with trauma survivors: Strengthening attachment bonds.* New York: Guilford Press.

Johnson, S. M. (2004). *The practice of emotionally focused marital therapy.* New York: Brunner-Routledge.

Kaptein, K. I., De Jonge, P., van den Brink, R. H. S., & Korf, J. (2006). Course of depressive symptoms after myocardial infarction and cardiac prognosis: A latent class analysis. *Psychosomatic Medicine, 68*, 662–668.

Karmilovich, S. E. (1994). Burden and stress associated with spousal caregiving for individuals with heart failure. *Progress in Cardiovascular Nursing, 9*(1), 33–38.

Kelley, H. H., & Thibaut, J. W. (1978). *Interpersonal relations: A theory of interdependence.* New York: Wiley.

King, K. K., & Reis, H. T. (2011). Marriage and long-term survival after coronary bypass surgery. *Health Psychology.* Published online August 2011.

Kloner, R. A., Mullin, S. H., Shook, T., Matthews, R., Mayeda, G., Burstein, S., et al. (2003). Erectile dysfunction in the cardiac patient: How common and should we treat? *Journal of Urology, 170*(2), S46–S50.

Kristofferzon, M.-L., Lofmark, R., & Carlsson, M. (2003). Myocardial infarction: Gender differences in coping and social support. *Journal of Advanced Nursing, 44*(4), 360–374.

Kulik, J. A., & Mahler, H. I. M. (2006). Marital quality predicts hospital stay following coronary artery bypass surgery for women but not men. *Social Science & Medicine, 63*, 2031–2040.

La Guardia, J. G., & Patrick, H. (2008). Self-determination theory as a fundamental theory of close relationships. *Canadian Psychology, 49*, 201–209.

La Guardia, J. G., & So, S. S. (2011). *Predictors of spouses' hypervigilance and control of patients in the aftermath of a cardiac crisis.* Unpublished manuscript.

Lane, D., Carroll, D., Ring, C., Beevers, D. G., & Lip, G. Y. H. (2002). The prevalence and persistence of depression and anxiety following myocardial infarction. *British Journal of Health Psychology, 7*(1), 11–21.

Lepore, S. J., & Coyne, J. C. (2006). Psychological interventions for distress in cancer patients: A review of reviews. *Annals of Behavioral Medicine, 32*, 85–92.

Lindau, S. T., Gosch, K., Abramsohn, E., Chan, P., Krumholz, H., Spatz, E., et al. (2010). *Gender differences in loss of sexual activity 1 year after an acute myocardial infarction (AMI).* Presentation at American Heart Association's Forum on Quality of Care and Outcomes Research in Cardiovascular Disease and Stroke, Washington, DC.

Linden, W., Phillips, M. J., & Leclerc, J. (2007). Psychological treatment of cardiac patients: A meta-analysis. *European Heart Journal, 28*(24), 2972–2984.

Lyons, R. R., Sullivan, M. J. L., Ritvo, P. G., & Coyne, J. C. (1995). *Relationships in chronic illness and disability.* Thousand Oaks, CA: Sage.

Mahrer-Imhof, R., Hoffmann, A., & Froelicher, E. S. (2007). Impact of cardiac disease on couples' relationships. *Journal of Advanced Nursing, 57*(5), 513–521.

Manne, S. L., Norton, T., Ostroff, J., Winkel, G., Fox, K., & Grana, G. (2007). Protective buffering and psychological distress among couples coping with breast cancer: The moderating role of relationship satisfaction. *Journal of Family Psychology, 21*, 380–388.

Manne, S. L., & Ostroff, J. S. (2008). *Coping with breast cancer: A couples-focused group intervention.* Therapist Guide. New York: Oxford University Press.

Manne, S. L., Ostroff, J. S., Norton, T., Fox, K., Goldstein, L., & Grana, G. (2006). Cancer-related relationship communication in couples coping with early stage breast cancer. *Psycho-oncology, 15*, 234–247.

Manne, S. L., Ostroff, J. S., Sherman, M., Heyman, R. E., Ross, S., & Fox, K. (2004). Couples' support-related communication, psychological distress, and relationship satisfaction among women with early stage breast cancer. *Journal of Consulting and Clinical Psychology, 72*(4), 660–670.

Manne, S., Ostroff, J. S., Winkel, G., Fox, K., Grana, G., Miller, E., et al. (2005). Couple-focused group intervention for women with early stage breast cancer. *Journal of Consulting and Clinical Psychology, 73*, 634–646.

Manne, S. L., Ostroff, J. S., Winkel, G., Grana, G., & Fox, K. (2005). Partner unsupportive responses, avoidant coping, and distress among women with early stage breast cancer: Patient and partner perspectives. *Health Psychology, 24*(6), 635–641.

Manne, S., Rubin, S., Edelson, M., Rosenblum, N., Bergman, C., Hernandez, E., et al. (2007). Coping and communication-enhancing intervention versus supportive counseling for women diagnosed with gynecological cancers. *Journal of Consulting and Clinical Psychology, 75*(4), 615–628.

Mauss, I. B., & Gross, J. J. (2004). Emotion suppression and cardiovascular disease: Is hiding feelings bad for your heart? In I. Nyklícek, L. Temoshok, & A. Vingerhoets (Eds.), *Emotion expression and health: Advances in theory, assessment and clinical applications.* Hove, UK: Brunner-Routledge.

Mikulincer, M., Florian, V., & Hirschberger, G. (2004). The terror of death and the quest for love: An existential perspective on close relationships. In J. Greenberg, S. L. Koole, & T. Pyszczynski (Eds.), *Handbook of experimental existential psychology* (pp. 287–304). New York: Guilford.

Moore, S. M. (1994). Psychologic distress of patients and their spouses after coronary artery bypass surgery. *AACN Clinical Issues in Critical Care Nursing, 5*(1), 59–65.

Moser, D. K., & Dracup, K. (2004). Role of spousal anxiety and depression in patients' psychosocial recovery after a cardiac event. *Psychosomatic Medicine, 66*, 527–532.

Newsom, J. T., & Schulz, R. (1998). Caregiving from the recipient's perspective: Negative reactions to being helped. *Health Psychology, 17*(2), 172–181.

O'Farrell, P., Murray, J., & Hotz, S. B. (2000). Psychologic distress among partners of patients undergoing cardiac rehabilitation. *Heart & Lung, 29*, 97–104.

Orth-Gomer, K., Wamala, S. P., Horsten, M., Schenck-Gustafsson, K., Schneiderman, N., & Mittleman, M. A. (2000). Marital stress worsens prognosis in women with coronary heart disease: The Stockholm Female Coronary Risk Study. *JAMA, 284*, 3008–3014.

Pennebaker, J. W. (1997). *Opening up: The healing power of expressing emotions.* New York: Guilford Press.

Rafaeli, E., & Gleason, M. E. J. (2009). Skilled support within intimate relationships. *Journal of Family Theory and Review, 1*, 20–37.

Randall, A. K., & Bodenmann, G. (2009). The role of stress on close relationships and marital satisfaction. *Clinical Psychology Review, 29*, 105–115.

Revenson, T. A., Abraido-Lanza, A. F., Majerovitz, S. D., & Jordan, C. (2005). Couples coping with chronic illness: What's gender got to do with it? In T. A. Revenson, K. Kayser, & G. Bodenmann (Eds.), *Couples coping with stress: Emerging perspectives on coping* (pp. 137–156). Washington, DC: American Psychological Association.

Robles, T. F., & Kiecolt-Glaser, J. K. (2003). The physiology of marriage: Pathways to health. *Physiology & Behavior, 79*(3), 409–416.

Rohrbaugh, M. J., Mehl, M. R., Shoham, V., Reilly, E. S., & Ewy, G. (2008). Prognostic significance of spouse "we-talk" in couples coping with heart failure. *Journal of Consulting and Clinical Psychology, 76*, 781–789.

Rohrbaugh, M. J., Shoham, V., & Coyne, J. C. (2006). Effect of marital quality on eight-year survival of patients with heart failure. *American Journal of Cardiology, 98*, 1069–1072.

Rollman, B. L., Belnap, B. H., LeMenager, M. S., Mazumdar, S., Houck, P. R., Counihan, P. J., et al. (2009). Telephone-delivered collaborative care for treating post-CABG depression: A randomized controlled trial. *JAMA, 302*(19), 2095–2103.

Rusbult, C. E., & Arriaga, X. B. (2000). Interdependence in personal relationships. In W. Ickes & S. Duck (Eds.), *The social psychology of personal relationships* (pp. 79–108). Chichester, UK: Wiley.

Ryan, R. M., & Deci, E. L. (2000). The darker and brighter sides of human existence: Basic psychological needs as a unifying concept. *Psychological Inquiry, 11*, 319–338.

Schulz, R., & Beach, S. R. (1999). Caregiving as a risk factor for mortality. *JAMA, 282*(23), 2215–2219.

Scott, J. L., Halford, W. K., & Ward, B. G. (2004). United we stand? The effects of a couple-coping intervention on adjustment to early stage breast or gynecological cancer. *Journal of Consulting and Clinical Psychology, 72*(6), 1122–1135.

Shemesh, E., Rudnick, A., Kaluski, E., Milovanov, O., Salah, A., Alon, D., et al. (2001). A prospective study of posttraumatic stress symptoms and nonadherence in survivors of a myocardial infarction (MI). *General Hospital Psychiatry, 23*, 215–222.

Shemesh, E., Yehuda, R., Milo, O., Dinur, I., Rudnick, A., Vered, Z., et al. (2004). Posttraumatic stress, nonadherence, and adverse outcome in survivors of a myocadial infarction. *Psychosomatic Medicine, 66,* 521–526.

Sher, T. G., & Baucom, D. H. (2001). Mending a broken heart: A couples approach to cardiac risk reduction. *Applied & Preventative Psychology, 10,* 125–133.

So, S. S., & La Guardia, J. G. (2011). Matters of the heart: Patients' adjustment to life following a cardiac crisis. *Psychology & Health, 26,* 83–100.

Strik, J., Denollet, J., Lousberg, R., & Honig, A. (2003). Comparing symptoms of depression and anxiety as predictors of cardiac events and increased health care consumption after myocardial infarction. *Journal of the American College of Cardiology, 42*(10), 1801–1807.

Suls, J., Green, P., Rose, G., Lounsbury, P., & Gordon, E. (1997). Hiding worries from one's partner: Associations between coping via protective buffering and distress in male postmyocardial infarction patients and their wives. *Journal of Behavioral Medicine, 20,* 333–349.

Theobald, K., & McMurray A. (2004). Coronary artery bypass graft surgery: Discharge planning for successful recovery. *Journal of Advanced Nursing, 47*(5), 483–491.

Thompson, D. R., & Cordle, C. J. (1988). Support of wives of myocardial infarction patients. *Journal of Advanced Nursing, 13,* 223–228.

Tucker, J. S. (2002). Health-related social control within older adults' relationships. *Journal of Gerontology, 57B:* 387–395.

Waltz, M. (1986). Marital context and post-infarction quality of life: Is it social support or is it something more? *Social Science Medicine, 22,* 791–805.

Waltz, M., Badura, B., Pfaff, H., & Schott, T. (1988). Marriage and the psychological consequences of a heart attack: A longitudinal study of adaptation to chronic illness after three years. *Social Science and Medicine, 27,* 149–158.

CHAPTER 11

Marital Dissolution and Physical Health Outcomes

A Review of Mechanisms[1]

DAVID A. SBARRA

University of Arizona

Marital separation is an unquestionably stressful event, even for adults who see divorce as an opportunity for personal growth and newfound freedoms. Routines are disrupted, work is missed, bank accounts are divided, lawyers are consulted, and furniture is moved—not to mention a long list of other chores. In terms of emotional and psychological experiences, it is arguable that there are few other events in life as *potentially* devastating as ending our closest relationships. For adults who initiate the separation and leave the marriage, guilt, uncertainty, and fear often commingle with a sense of excitement over the potential relief of ending an unfulfilling relationship. For those who are left by their partners, rejection, loneliness, shame, and longing can evolve into clinical depression, prolonged grief, and substance abuse. In marriages with children and two invested parents, separation means dealing with the fact that you often cannot see your children as much as you or they would like, which, in turn, can fuel a deep sense of emptiness and loss.

[1] The author's preparation of this chapter was supported by awards from the NIH/NIA (R21 AG#028454), NIH/NIMH (R03 MH#074637), and from NSF (BCS #0919525). Mark Borgstrom at the University of Arizona provided technical assistance with the forest plot presented in Figure 11.1.

Research substantiates these claims. Marital separation/divorce are associated with increased rates of both poor mental and physical health outcomes (Amato, 2000; Kiecolt-Glaser, Fisher, Ogrocki, Stout, Speicher, & Glaser, 1987; Sbarra & Nietert, 2009; Williams & Umberson, 2004), including death by suicide and homicide (Kposowa, 2000). Despite these risks, most divorced adults can be described as resilient or recovered, with the former group evidencing few disruptions in well-being during the separation and the latter group evidencing a distinct period of improvement after the separation. For example, in a recent panel study of over 16,000 German adults and over 600 divorces, Mancini, Bonanno, and Clark (2009) found that about 71% of divorcing adults report little change in subjective well-being from 4 years prior to and up to 4 years postdivorce, about 9% move from low levels of well-being before the divorce to high levels of well-being after the divorce, and about 19% move from moderate levels of well-being before the divorce to low levels of well-being after the divorce; this latter group constitutes those people who are most adversely impacted by the end of marriage—they experience decrements in well-being and do not recover to the predivorce functioning in the sequent 4 years. These findings are consistent with other work demonstrating that resilience in the face of divorce and other stressful life events is the norm, not the exception (Amato; Bonanno, 2004; Hetherington & Kelly, 2002; Sbarra & Emery, 2005).

The unfortunate and sometimes unrecognized side of the resilience story is that because divorce is such a common event, 10–20% of people experiencing precipitous drops in their well-being and quality of life (following a separation experience) amounts to a large number of people for whom this life event has lasting negative effects. The incidence of divorce in the United States has decreased over the last 35 years, from 5.0/1,000 marriages in 1970 to 3.5/1,000 in 2008 (Bramlett & Mosher, 2002; Tejada-Vera & Sutton, 2009). Using data from the National Survey of Family Growth, which involved face-to-face interviews with 10,847 women ages 15 to 44 years, Bramlett and Mosher estimated the divorce rate for first marriages in the United States to be 43%. Among adults who do divorce, 15% remarry with 1 year, 39% within 3 years, and 75% with 10 years; close to 40% of second marriages break up within 5 years (Bramlett & Mosher). Given these estimates, it is critical to better understand the variables and processes that help explain why marital separation is an acute, transient stressor for some adults but can evolve into chronic stressor for others (cf. Amato, 2000).

This chapter focuses on *how* the stress of divorce can be associated with physical health outcomes. I limit the focus of this review to literature on physical (relative to mental) health outcomes for two primary reasons. First, there is increasing evidence that marital separations are associated with poor physical health outcomes, yet the bulk of the findings in this area, which often reside in fields outside of psychology, are not reviewed as a coherent body of research in a single space. This chapter attempts to provide some order to the literature. Second, no reviews have outlined the different mechanisms of action that may link marital separation to health-related outcomes, and doing so helps bring into relief questions that must be answered before the study of divorce-related coping can be significantly advanced. Similarly, the review is limited to marital separation/divorce because recent analyses have investigated health outcomes related to bereavement (Stroebe, Schut, & Stroebe, 2007). For example, a recent meta-analysis of over 250,000 adults found that the relative risk associated with early death was significantly elevated among widowed adults relative to their married counterparts (Manzoli, Villari, Pirone, & Boccia, 2007). Clearly, some of the mechanisms discussed herein—such as biological reactivity associated with the loss-induced negative affect—apply equally well to the study of health outcomes following both divorce and bereavement. In other instances, however, it is clear that there are aspects of marital separation that may be uniquely associated with end-point health outcomes. Thus, given the space constraints of the chapter, I focus only on marital separation/divorce in an effort to highlight the health-relevant mechanisms that unfold when marriage ends by choice rather than through the death of a spouse.

The chapter has five main sections. I provide a brief theoretical overview discussing why social separations like divorce may be a relatively unique type of stressful event that has the potential to disrupt health outcomes. Second, I review epidemiological literature linking marital separation/divorce to all-cause mortality. Establishing the size of macrolevel associations sets the stage for understanding mechanisms of action. The remaining sections focus on reviewing what we know and what we do not know about putative mechanisms by considering four areas of research: (1) evidence for social selection processes (whereby third variables explain the association with both divorce and increased risk of illness), (2) structural changes (e.g., changes in income among single parents), (3) health behavior changes, and (4) emotional and stress-related changes. From the outset, it is important to recognize that these pathways are not orthogonal. For example, hostility may serve as a third-variable explanation for both divorce and early mortality, and this dimension of personality also is associated with health outcomes by virtue of changes in health behaviors and emotional reactivity, which can spur cardiovascular reactivity and prolong physiological stress responses (Miller, Smith, Turner, Guijarro, & Hallet, 1996).

From Coregulation to Dysregulation

Are relationship disruptions uniquely stressful? In this time of nearly unprecedented financial calamity, it is reasonable to argue that divorce is just one among a long list of difficult life experiences, including sustained unemployment or losing a home to foreclosure. Although the literature contains no direct comparisons of the differential stress associated with divorce and something like unemployment, it is long recognized that social disruptions are among life's most stressful events. In Holmes and Rahe's (1967) now-classic effort to hierarchically organize events that necessitate social readjustment, divorce and marital separation were sandwiched between the death of a spouse and a jail term among the first four most difficult life events. Mounting empirical evidence suggests it is not a coincidence that the unwanted end of social relationships is so distressing. Sbarra and Hazan (2008) reviewed literature on the psychobiology of attachment to propose an integrative coregulation–dysregulation model (CDM) of human loss experiences. A primary hypothesis of this model is that mammalian loss experiences disrupt the biological systems that subserve the psychological experience of felt security. From this perspective, the well-documented biological and psychological dysregulation people experience upon unwanted separations is a consequence of removing the functional components of the attachment system (cf. Shear & Shair, 2005). When this occurs, the biological systems that regulate and are regulated by a relationship run free, which can lead to a number of undesirable end points.

One of the main implications of the CDM model is that psychological reactions to the separation experience can prolong biological dysregulation. Of course, this is the case for any stressful event; how people appraise difficult events has distinct biological correlates (e.g., Tomaka, Blascovich, Kelsey, & Leitten, 1993), and psychological reactions that promote biological reactivity are believed are harmful to health (Dickerson & Kemeny, 2004; Kemeny, 2003; Kiecolt-Glaser, McGuire, Robles, & Glaser, 2002). With loss experiences, the CDM holds that the disruption of coregulation, which is defined as the reciprocal maintenance of psychophysiological homeostasis within a relationship, engenders a biological response akin to a withdrawal reaction. How adults regulate this state has important implications for their long-term health. One way of suppressing physiological dysregulation in the aftermath of a separation is with alcohol and other substances (Agrawal & Lynskey, 2009; Dawson, Grant, Stinson, & Chou, 2006). Alternatively, people can succeed or fail in creating meaning from the event, and this process has clear health correlates. For example, in a study of 40 HIV-positive men following the loss of their partner from AIDS, cognitive processing that resulted in finding meaning from the loss experience proved protective to immune system declines that index the emergence of full-blown AIDS (Bower, Kemeny, Taylor, & Fahey, 1998). Most interestingly, this study demonstrated that cognitive processing in the *absence* of finding meaning was associated with a significant drop in immune functioning over time; repetitively processing loss-related thoughts without coming to some type of larger realization or insight is correlated with distinct biological risk. These simple examples suggest that how people regulate felt security in the aftermath of a loss experience is associated with long-term health outcomes. As described herein, these so-called regulatory pathways (i.e., changes in health behaviors and changes in psychological appraisals of the situation) operate hand in hand with structural changes and selection factors to protect against or increase risk for health problems following divorce or separation experiences. Because disruptions in felt security are a core feature of relationship disruptions like divorce, there is reason to believe that these events—relative to other potentially negative life events—have relatively unique potential to activate biological profiles that are consistent with threat appraisals. Once induced, it is this biological dysregulation that must be quelled in the process of recovery. As I describe here, some modes of coping with marital separation/divorce are ill-suited to helping people restore a sense of felt security.

Broad-Based Health Outcomes: Documenting Effects of Interest

The empirical study of the association between marital status and early all-cause mortality dates back well over a century (Farr, 1858). This section of the chapter briefly reviews the psychological, sociological, and epidemiological research examining the association between divorce–marital separation and risk for early mortality from all causes. A systematic and quantitative review of this literature is needed for two chief reasons. First, null findings exist, and, without synthetic reviews, there is no way to determine if the association between divorce and later health is reliably different from zero. Said differently, there is no need to study mechanisms if the field cannot conclude, without dispute, that divorce is a clear public health concern. Second, as noted already, the research in this area is disparate, and summating across different literatures is the only way to achieve an accurate appraisal of the problem. Unfortunately, the space constraints of this chapter preclude an exhaustive assessment of the quantitative association between divorce and early mortality as well as increased morbidity among the ill.

I limit my review to prospective empirical studies that report results from Cox's proportional hazard model (i.e., a survival analysis) and that include confidence intervals on the parameter estimates. I used these confidence intervals to derive the standard error (SE) of measurement for the estimate, and from the SE I applied an inverse variance procedure to weight the estimate from a given study by the SE, which results in large studies (with more precise effect size estimates) being given a larger weight in the meta-analysis (Higgins & Green, 2009). The effect size of interest in the reports reviewed here is the proportional risk hazard (RH) statistic. Similar to the relative risk statistic, the RH is the ratio of the proportion of separated or divorced adults who die early among all separated or divorced adults relative to the proportion of married adults who die early among all married adults (the risk of those adults exposed to separation/divorce divided by the risk of the unexposed); the RH is interpreted like the RR, with values greater than 1.0 signifying a percentage of increase in risk at any given study period for adults exposed to divorce.

Figure 11.1 displays summary data from 12 studies (Ben-Shlomo, Smith, & Shipley, 1993; Cheung, 2000; Dupre, Beck, & Meadows, 2009; Ebrahim, Wannamethee, McCallum, Walker, & Shaper, 1995; Ikeda et al., 2007; Iwasaki et al., 2002; Johnson, Backlund, Sorlie, & Loveless, 2000; Lund, Christensen, Holstein, Due, & Osler, 2006; Malyutina, Bobak, Simonova, Gafarov, Nikitin, & Marmot, 2004; Mendes De Leon, Appels, Otten, & Schouten, 1992; Molloy, Stamatakis, Randall, & Hamer, 2009; Nilsson, Nilsson, Ostergren, & Berglund, 2005), with 29 independent effects, assessing risk for early death following more than 25,000 marital separation experiences in a total combined sample of 366,610 adults from eight different countries. The average study in this sample followed participants 147 months, with the longest follow-up period being 216 months and the shortest being 84 months. As shown in the figure, the confidence interval of the summary RH statistic does not cross zero. Thus, it can be concluded with 95% certainty that the average RH emerging from this sample of prospective studies falls between 1.43 and 1.50, indicating that separated or divorced adults were between 43 and 50% more likely to be dead at each follow-up assessment than adults who were married at the inception of the study. Studies with multiple effects (e.g., Johnson et al., 2000) report RH statistics for different subgroups of interest; the variables (e.g., participant sex or ethnicity) can be thought of as potential moderators, but the total number of available samples precludes determining if the average RH is different across groups using a fixed-effect model. In summarizing these findings, it is important to note that the average RH is derived from Cox models that adjusted only for age. A subsample of the papers in Figure 11.1 report adjusted models in which the authors attempted to eliminate effects of marital status via statistical control (e.g., tobacco use, baseline blood pressure, income, labor force status, body mass index). Among the 17 adjusted effect sizes, the average RH was 1.32 with a weighted confidence band of 1.31 to 1.33. Thus, we can conclude with certainty that divorced or separated adults evidence increased rates of early, all-cause mortality even in the fully adjusted models.

The results of the present analysis are consistent with prior work in this area of study (Hu & Goldman, 1990). Nonetheless, although meta-analysis of epidemiological research is useful for establishing the average association between separation/divorce and later health outcomes, many

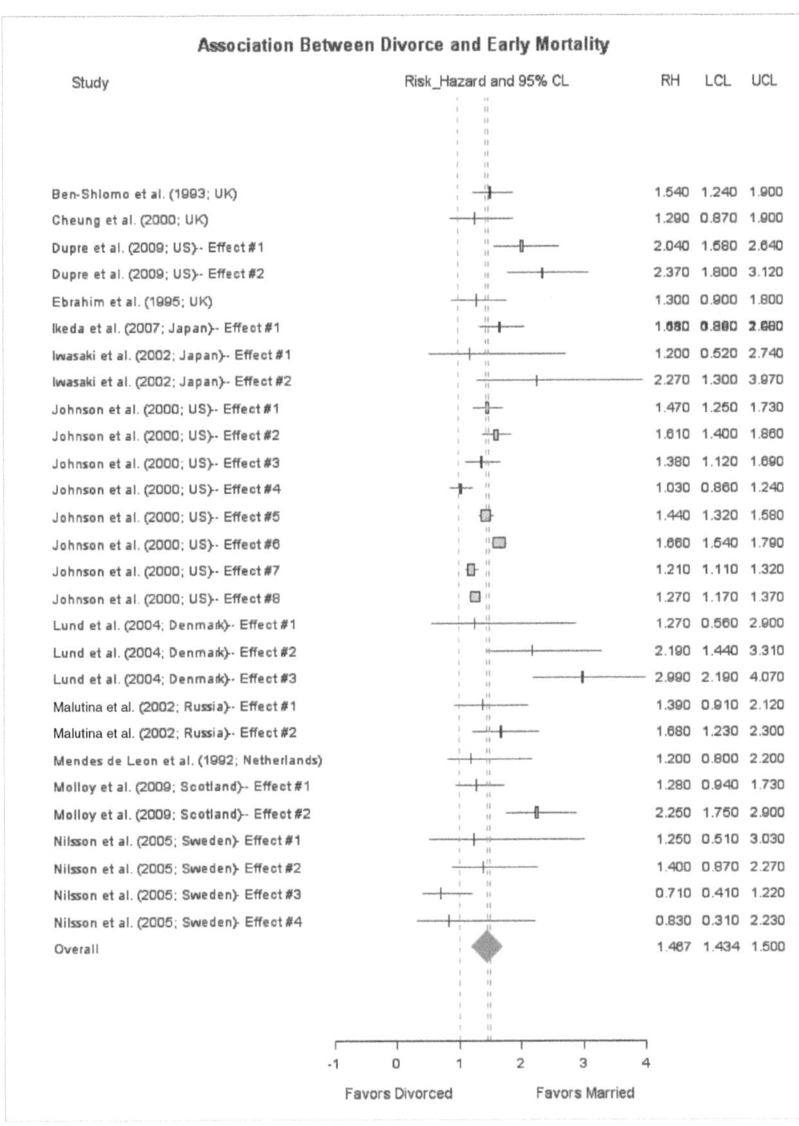

Figure 11.1 Forest plot depicting mean risk hazard (RH) effect sizes from 12 different studies examining the prospective association between marital separation/divorce and risk for early mortality. CL = confidence limit (LCL = lower bound of 95% CL; UCL = upper bound of the 95% CL). Each effect size is weighted using an inverse variable approach, and this weighting is reflected in the size of effect size estimate as well as the confidence band surrounding the effect. The single dashed line represents the unity, and confidence intervals that cross the single dashed line are not statistically significant. The double dashed line represents the confidence interval for the average weighted effect size across all RH statistics ($N = 29$) included in the meta-analysis.

epidemiological studies present a relatively gross and static picture of marital status. Several studies suggest an important series of qualifications to the meta-analytic findings. For example, in a 40-year follow-up study of separated or divorced adults, Sbarra and Nietert (2009) recently reported that the risk associated with being a separated or divorced adult was reduced to zero after changes in marital status were taken into account; specifically, adults who remained separated throughout the follow-up period were the harbingers of risk, whereas adults who separated/divorced from a partner but later remarried evidenced no greater risk for early death than adults who remained married throughout adult life (cf. Dupre et al., 2009). Given these findings, the literature in this area must continue to be viewed as nascent until a more precise picture of the moderators of risk emerges.

Mechanisms Linking Separation/Divorce and Health: Social Selection

The social selection perspective holds that some people possess characteristics that increase risk for both separation/divorce and poor health outcomes (Amato, 2000; Goldman, 1993; Osler, McGue, Lund, & Christensen, 2008; Wade & Pevalin, 2004). From this point of view, the outcomes thought to be associated with marital dissolution are mere third-variable correlates. This view stands in contrast to a social causation perspective, whereby health problems are believed to be a consequence of divorce-related processes (e.g., changes in health behaviors, loss of income, chronic stress). The social selection perspective that separation/divorce outcomes are nonrandom is supported by the well-known existence of gene-by-environment correlations (i.e., the observation that environments are shaped by genotypes; Scarr & McCartney, 1983) and by evidence from the behavior genetic study of marital dissolution. McGue and Lykken (1992) found greater concordance of divorce in monozygotic than dyzygotic twins, indicating that a substantial portion of the risk for marital dissolution is explained by genetic factors. A follow-up study demonstrated that between 30% (in women) and 42% (in men) of heritable divorce risk was attributable to personality differences (Jockin, McGue, & Lykken, 1996), and other work has shown that genetic association with controllable life events, of which divorce is one, is entirely explained by differences in personality (Saudino, Pedersen, Lichtenstein, McClearn, & Plomin, 1997). Consistent with these findings, Whisman, Tolejko, and Chatav (2007) examined the association between diagnosed personality disorders and the probability of marital disruption in a sample of over 43,000 adults in the United States, finding that all personality disorders were associated with a significantly increased risk for divorce. Furthermore, there is clear evidence that health behaviors such as alcohol and drug use predict the future likelihood of divorce (Fu & Goldman, 2000), which suggests that these problems reliably select adults out of marriage.

Regrettably, very few studies exist that link divorce and health via personality, health behaviors prior to the divorce, or other heritable mechanisms. Using data from the Terman Life Cycle study, for example, Tucker, Friedman, Wingard, and Schwartz (1996) reported that the risk associated with having ever experienced a divorce and early mortality could be reduced (by 21% for men and 15% for women) after accounting for childhood conscientiousness and a history of parental divorce. This paper aside, documenting the existence of third-variable selection processes is often one of cobbling together established associations; if divorce risk can be explained by personality, and there is a voluminous literature linking personality and health outcomes (G. E. Miller, Cohen, Rabin, Skoner, & Doyle, 1999; Roberts, Kuncel, Shiner, Caspi, & Goldberg, 2007), working hypotheses must include personality attributes as those that predict both the end of marriage and poor health outcomes.

Epidemiological research seeking to account for selection effects when investigating differences in health between married and divorced people has revealed mixed results. In a sample of over 3,500 British adults, Cheung (1998) examined the social class, education, health, and mental health predictors of divorce 10 years later; the significant predictors of divorce were then entered into models examining differences between the divorced and nondivorced. The predictors of divorce eliminated small differences between married and divorced men but not differences between married and divorced women (Cheung). This pattern of results is consistent with the adjusted meta-analytic findings previously described. One methodological problem in disentangling findings in this area is that investigators often lump pre- and post-separation/divorce variables together when seeking to (statistically) eliminate differences between married and divorced adults. Consequently, it is difficult to determine what variables are hypothesized to *predict* the separation/divorce and what variables are hypothesized to be a *consequence* of the separation/divorce. Improved measurement using time-varying indicators (e.g., self-rated health prior to and following a separation; see Dupre et al., 2009) will address this confusion in the literature.

Overall, it can be concluded that selection effects do operate to predict the likelihood of future divorce, but these effects cannot explain the entirety of the separation/divorce–health association. The most likely third-variable explanation for the separation/divorce–health association is personality and temperamental differences, which are well-established predictors of the genetic variance associated with marital dissolution. Finally, it is important to note that powerful models exist for disentangling the genetic selection and life event causation effects for outcomes of interest. Osler and colleagues (2008) recently used a co-twin control design to investigate rates of health outcomes between twins who were discordant for widowhood or divorce. The results indicated that depression and rates of smoking may be consequences of ending a marriage, but differences in many other health outcomes (e.g., self-rated health, alcohol use, body mass index [BMI]) may be due to underlying genetic explanations, not the stress of a relationship transition. One additional approach for investigating selection and causation effects is the children of twins model (D'Onofrio et al., 2003), which allows for a quantification of the environmental and genetic processes underlying the association between marital disruption and health outcomes. Until this model can be applied to the study of separation/divorce-health associations, disentangling selection and causation effects will continue to be difficult.

Mechanisms Linking Separation/Divorce and Health: Structural Changes

I define structural variables as macrolevel processes that alter the context of how adults may respond to the end of marriage. Changes in these structural variables can moderate processes that are more proximally related to divorce–health associations by maintaining states of chronic stress (see Amato, 2000). The first structural change I consider is financial—divorce is a costly life event, both in terms of legally ending a marriage as well as the long-term consequences for a family's financial health. Whereas evidence exists that divorce has a negative impact on both men and women's cumulative earning power over the life course (Waite & Gallagher, 2000; Wilmoth & Koso, 2002), there is perhaps no finding in the divorce literature as consistent as the fact that women, especially those who become single mothers, experience precipitous drops in their financial well-being (Amato, 2000; Peterson, 1996; Weitzman, 1985). Peterson operationalized adults' standard of living as the ratio of their income to needs (I:N) and then reported that, in the year following a divorce, women evidenced a 27% *decrease* in their I:N levels, whereas men evidenced a 10% *increase* in their I:N levels. Although the data from this study are now 30 years old, more recent investigations, including those with cross-national data, corroborate the basic finding of major changes in women's economic well-being following a divorce (Uunk, 2004; Zagorsky, 2005).

Although there exist no direct data implicating changes in finances as a complete explanation for the separation/divorce–health association, evidence does indicate that differences in socioeconomic status between married and divorced adults can partially account for differences in all-cause mortality. Using data from the U.S. Health and Retirement Study, Dupre et al. (2009) found that accounting for socioeconomic differences (indexed by length of time in a specific job, percentage of group members in a professional or managerial occupation, overall wealth, and health insurance coverage) reduced the all-cause mortality disparity between married and divorced adults by 24% in women and by 35% in men. Interestingly, this study also considered the effects of behavioral risk factors (e.g., alcohol consumption, tobacco use, exercise patterns) and health factors (e.g., chronic diseases, limitation in daily activities, and depression) on reducing mortality differences between married and divorced participants; for both men and women, socioeconomic changes resulted in equivalent or larger reductions in risk between the groups than the behavioral or health factors (Dupre et al.), suggesting that structural change in financial well-being is a robust predictor of important health outcomes. It is widely believed that socioeconomic status exerts a deleterious influence on health via chronic psychological stress and changes in health behaviors (Adler et al., 1994). The next two sections of the chapter consider how these more proximal processes operate following separation/divorce.

In addition to changes in finances, the size and diversity of social networks change dramatically following marital separation and divorce (Milardo, 1987). Friendships are often divided and recast, and family relationships need to be redefined. Rands (1988) reported that 40% of people lose contact with members of their social network upon becoming divorced and that social networks become smaller and less dense. As a structural variable, network density is highly associated with health outcomes (Berkman, Glass, Brissette, & Seeman, 2000; Ertel, Glymour, & Berkman, 2009; House, Landis, & Umberson, 1988). For example, in a now-seminal study, House et al. reported a strong linear association between level of social integration and mortality risk across several large-scale epidemiological studies. A recent study of over 2,300 Dutch adults provides new information about the specific ways that social networks are disrupted following a divorce (Kalmijn & van Groenou, 2005). Although divorced adults were found to spend more time with friends than their married counterparts, they also had less neighborhood contact, a decrease in church attendance and club participation (for women only), and a decrease in outdoor recreation (for men only). One of the chief findings from this study is the identification of a potential mediating process, whereby the statistical effect of divorced and married women's social integration was explained by the availability of contextual resources, including living in poorer neighborhoods, having fewer financial resources, and having less available time to participate in social activities (Kalmijn & von Groenou).

Overall, there is a fair amount of evidence suggesting that structural changes—operationalized here as changes in financial resources and social networks—play a role in explaining the association between separation/divorce and physical health outcomes. Divorce exerts a large negative effect on the accumulation of wealth and has an especially negative effect on financial resources for women and single mothers. Epidemiological data suggest that changes in financial resources alone play a large part in explaining why divorced adults exhibit worse health outcomes than their married counterparts. Beyond finances, the dissolution of marriage also exerts a negative effect on social network integration. Although no studies have examined if changes in social network size or density explain the separation/divorce–health association, a large literature documents that these variables are of great importance when considering risk for early mortality.

Mechanisms Linking Separation/Divorce and Health: Health Behaviors

The phrase *health behaviors* is an umbrella description for the actions people take to improve or worsen their overall health. Salubrious health behaviors include participating in regular exercise, maintaining a healthy weight, consuming a balanced diet, and getting restorative sleep on a routine basis; unhealthy behaviors include tobacco use, excessive alcohol use, substance abuse/dependence, and high-risk sexual activity. With respect to the association between separation/divorce and risk for early mortality, a strong hypothesis is that the stress of marital dissolution coupled with the loss of social control of health behaviors leads to a decrease in salubrious activities and an increase in unhealthy activities, and it is these changes that can explain why separated/divorced adults are at increased risk for early mortality. The social control model of health behaviors refers to the idea that "relationships may provide social control of health behaviors indirectly by affecting the internalization of norms for healthful behavior, and directly by providing informal sanctions for deviating from behavior conducive to health" (Umberson, 1987, p. 309). When marriages dissolve, social control changes dramatically; married adults report that their spouse is the most frequent person to remind them about their health and health behaviors, whereas separated/divorced adults report that the most frequent reminders for good health come from unrelated adults in their life (Umberson, 1992).

Marital separation and divorce are associated with a wide range of negative health behaviors. Cross-national epidemiological data indicate that, as a demographic variable, being separated/divorced is associated with substantially elevated risk for severe insomnia and problems of sleep maintenance (Doi, Minowa, Okawa, & Uchiyama, 2000; Hajak, 2001). In a 3-year follow-up study of over 30,000 adults, becoming separated/divorced was significantly associated with the first-time initiation of cannabis use in previous abstainers (Agrawal & Lynskey, 2009). Among adults diagnosed with prior to past-year alcohol dependence, ending a first-marriage is associated with an increased risk for nonabstinent recovery (Dawson et al., 2006), and women's rate of alcohol consumption increases immediately prior to a marital separation (Mastekaasa, 1997). In epidemiological research, Joung, Stronks, Van de Mheen, and Mackenbach (1995) found that a set of health behavior variables (including tobacco use, alcohol consumption, eating a regular breakfast, rates of exercise, and BMI) significantly reduced differences between married and divorced adults' perceived general health and subjective complaints as well as the difference in the number of women's chronic medical conditions. The set of health behaviors led to an 11% reduction in the difference between married and divorced men's ratings of subjective health and a 14% reduction in the difference between married and divorced women's ratings of subjective health. Ikeda and colleagues (2007) conducted a similar study on objective risk for early mortality and found that a set of health behavior predictors reduced but did not eliminate the association between divorce and risk for early mortality in men; in this sample of Japanese adults, divorced women were not at elevated risk for early mortality relative to their married counterparts.

One of the limitations of these studies is that they are not time based and, consequently, do not reveal how changes in marital status are associated with changes in health behaviors. Two studies provide more rigorous data on this topic. In a sample of over 80,000 US women from the Nurse's Health Study, Lee, Cho, Grodstein, Kawachi, Hu, and Colditz (2005) reported that women who became divorced over a 4-year period evidenced significant decreases in their BMI (relative to women who remained married), evidenced a significant likelihood of smoking relapse (among prior smokers who were abstaining at the start of the study), and evidenced a significant decrease in fruit and vegetable consumption. Two elements of these findings are noteworthy. First, the effects are multivariate adjusted; thus, the smoking relapse effect is robust and remained significant even when accounting for a variety of other competing predictors. Because the analyses are derived from a rigorous prospective study, these results imply that smoking relapse is associated with the stress that surrounds divorce. Second, it is important to recognize that the findings are limited to these three outcome variables and not all aspects of health behavior. For example, no significant differences were observed between married and divorced women's alcohol consumption or rates of smoking initiation among nonsmokers (Lee et al.). For women, the best available data suggest that divorce does not have a sweeping impact on all health behaviors. Findings from a parallel study of men's health behavior in the wake of divorce suggest a similar conclusion. Like women, men who became divorced evidenced a significant decline in their BMI, and they also evidenced a general decrease in daily activity levels (Eng, Kawachi, Fitzmaurice, & Rimm, 2005). Divorced men also evidenced a significant decrease in fruit and vegetable consumption and reported higher rates of alcohol consumption than married men. For the latter effect, *change* in alcohol consumption was not significantly different between those men who remained married and those who became divorced; thus, it is possible that alcohol consumption differences predated the end of the marriage.

Overall, one of the strongest conclusions that can be drawn from the study of health behaviors among divorced adults is that there is not nearly enough research on the topic. The two previously described epidemiological studies are the only ones to document how changes in marital states are associated with *changes* in health behaviors, and these investigations, although thorough and based on exceptionally large samples, did not include assessments of sleep changes or drug use. Despite these limitations, there is fairly clear evidence that, for women at least, the end of marriage is associated with smoking relapse. This area of study is dominated by epidemiological research and would benefit from psychological studies designed to assess how the psychological stress of divorce is associated with changes in health behaviors. Another way of saying this is that the field will benefit from studies using smaller samples with higher-quality measurement of the constructs of interest.

Mechanisms Linking Separation/Divorce and Health: Psychological Stress and Emotional Dysregulation

The last 15 years have witnessed a surge of research evidence demonstrating that psychological stress is correlated with disruptions in a range of biological systems, including the autonomic nervous system, the neuroendocrine system, and the immune system, as well as a range of disease processes (Cohen & Herbert, 1996; Cohen & Rodriguez, 1995; Kiecolt-Glaser et al., 2002; Miller & Blackwell, 2006; Sapolsky, 1996, 1998; Segerstrom & Miller, 2004). Only a handful of studies have examined how the psychological responses to marital separation may be associated with biomarkers that have distinct health implications. The work in this area began in the 1980s with a series of now seminal studies by Kiecolt-Glaser and colleagues (Kiecolt-Glaser et al., 1987, 1988). In the first report of this series, women who were recently separated from their spouse (within the past year) evidenced poorer immune functioning than sociodemographically matched married women. Specifically, recently separated women demonstrated poorer immune proliferation in response to two mitogens, significantly lower percentages of Natural Killer (NK) cells, and significantly higher antibody titers to the Epstein Barr Virus (EBV), which serves as an indirect measure of cellular immune responses. Within the separated/divorced sample, the degree of attachment to an ex-partner also was associated with greater impairments in immune responding (Kiecolt-Glaser et al., 1987). In a separate study, similar results were observed for men, although immunological differences between married and separated/divorced men were not as pronounced as observed in women. Separated/divorced men evidenced significantly higher functional immune parameters, including elevated antibodies to EBV and herpes simplex virus (HSV). Among the separated/divorced men, those who initiated the separation reported better health and evidenced better antibody titer responses to EBV (Kiecolt-Glaser et al., 1988).

After Kielcolt-Glaser and colleagues published their findings on immune responses following separation/divorce, research investigating how the stress of marital dissolution may be correlated with health-relevant biomarkers came to a virtual standstill.[2] Since the late 1980s, only two studies have been conducted in this area. Powell and colleagues (Powell et al., 2002) used divorce as a model of chronic stress in middle-aged women and examined neuroendocrine and catecholamine responses compared to a matched sample of nonstressed women. The findings revealed that divorced women evidenced significantly elevated cortisol responses in the evening relative to nondivorced women; surprisingly, divorced women also evidenced elevated rates of testosterone, which the authors interpreted as a possible effect of experiencing an emotional release from a poor marriage (cf. Mazur & Michalek, 1998). The findings from this study thus revealed a mixed picture for the health of divorced women.

More recently, Sbarra and colleagues (Sbarra, Law, Lee, & Mason, 2009) designed a laboratory-based study of autonomic nervous system responses to test the idea that physiological reactivity following marital separation is correlated with the degree of emotion regulatory effort adults need to invoke when thinking about their separation experience. The overarching theory behind this research is that emotion regulatory responses (e.g., catastrophizing an event in your mind, or reappraising an event as benign and personally significant) serve as the mediators connecting specific individual difference variables with poor divorce-related outcomes; for example, one hypothesis consistent with this idea is that people with a high degree of attachment anxiety show greater physiological responses to thinking about their separation experience because they view (i.e., appraise) the separation as a chronically threatening event for their well-being. Using

[2] One possible explanation for why research in this area slowed quite dramatically is that (in the 1990s) researchers in psychoneuroimmunology (PNI) focused a great deal of effort creating research paradigms that could interrogate clearly defined biological endpoints. In the study of relationships, a wound healing paradigm was introduced (e.g., Kiecolt-Glaser, Marucha, Malarkey, Mercado, & Glaser, 1995; Robles, 2007) that had clear relevance to clinical health outcomes. These models dominated the study of close relationships for a long period of time, and it is only recently, as the PNI and health psychology fields have before more refined, that investigators have once again cast the study of relationships more broadly.

these ideas as a basis for their study, Sbarra et al. (2009) found that participants who reported greater divorce-related emotional intrusion (e.g., dreaming about the separation, experiencing waves of sudden emotion about the separation) entered the study with significantly higher levels of resting systolic and diastolic blood pressure (BP). In addition, during a task in which participants mentally reflected on their separation experience, men who reported that the task required a great deal of emotion regulatory effort (i.e., feeling upset combined with a need to exert control of one's emotions to prevent a worsening of distress) evidenced the largest increases in BP, and these effects were in addition to those observed for baseline functioning (Sbarra et al.). One important aspect of this study was that the observed changes in blood pressure among highly distressed men placed them in a hypertensive range (systolic BP > 150 mg/mm) that would typically warrant medical intervention if observed as resting BP. Although many activities elevate BP into this range on a temporary basis, it is plausible that one way divorce may exert health compromising effects is through emotion regulatory demands; if highly distressed men repeatedly push their BP responses into the range of hypertensive functioning, this would be of clinical significance. This idea is consistent with current models of allostatic load (McEwen, 1998), in which psychological stress can exert its effects on health by taxing regulatory systems (e.g., the vascular system) through repeated "hits" of high activity or a failure to shut down normative responses to stress.

Overall, only a small number of studies exist linking the psychological stress of separation/divorce with biological responses that are health relevant. To fully understand the mechanisms connecting marital dissolution and broad-based health outcomes, many more studies of this nature are needed. Important moderator variables, including attachment style, personality, and divorce-related variables (e.g., who initiated the separation), can be introduced to this line of research; presently, very few of these variables are studied in relation to health outcomes following divorce. Although the four studies previously reviewed are clear in demonstrating that the psychological stress of separation/divorce can be associated with a range of different biomarkers, it is important to recognize that these effects may better explained by social selection than as a consequence of separation/divorce. If personality both selects adults out of marriage and predisposes people to respond a certain way to the separation (e.g., with greater emotional reactivity in a laboratory paradigm), then it cannot be concluded that emotional reactivity is causal in the pathway from marital separation to health outcomes. This methodological problem can be addressed through careful experimentation. Prospective intervention studies targeting adults' psychological appraisals of the separation/divorce experience can be designed to assess the potential mediating mechanisms linking intervention strategies with important health outcomes. This call to introduce experimentation into this study of divorce adjustment is consistent with a recent plea for increased experimentation in the area of social relationships and health outcomes; Cohen and Janicki-Deverts (2009) argue that nearly all of the literature—from psychology to epidemiology—linking social relationships and health outcomes is correlational. Until small-scale experiments and large-scale randomized controlled trials of interventions to prevent divorce-related distress are implemented, the field will remain without a detailed account of the *causal* mechanisms of action linking marital dissolution with health outcomes.

Concluding Remarks and Future Directions

In this chapter, I provided quantitative data evaluating the strength of the association between marital separation/divorce and risk for early death by all causes. The results of a small meta-analysis examining the magnitude of the RH indexing this association revealed a reliable and statistically significant difference in early mortality between married and separated/divorced adults. Using these findings as a starting point, I then reviewed literature concerning four potential mechanisms of action: (1) social selection processes, (2) changes in structural resources and social networks, (3) changes in health behaviors, and (4) the association between emotional reactivity and biomarkers with clear health relevance.

One of the most pressing questions for this area of study is reconciling two disparate observations. On one hand, there is clear evidence that most adults are resilient in the face of their divorce. Many people experience few lasting disruptions to their overall well-being when their marriage ends (Mancini et al., 2009), and this observation is consistent with the general human capacity to thrive when confronting difficult life experiences (Bonanno, 2004). On the other hand, the meta-analytic data presented here are fairly clear in suggesting that separation/divorce is associated with increased risk for all-cause mortality. Taken together, the resilience and risk observations paint rather different pictures about the impact of divorce on health outcomes. One way to understand this difference is statistical. Resilience is the most common outcome following divorce, but it is incorrect to equate the modal response with the mean response. In terms of the mortality effect, this is a mean response and is susceptible to the influence of outliers, which, in this case, would be the small percentage of adults who have extreme and chronic difficulties when their marriage ends. Distinctions of this nature are worth making because they suggest that we continue to search for the key moderating variables that distinguish between those adults who exhibit early patterns of resilience and those who exhibit lasting difficulties. With respect to the latter group, we have identified very few variables that are associated with biomarkers that have clear-cut relevance for health outcomes (e.g., blood pressure reactivity, inflammation, BMI).

Given these comments, one way to make advances in the field is to incorporate more complete assessments of moderating variables into the study of the four mechanistic pathways described already. Aside from personality characteristics, are there other variables that predict increased risk for both divorce and poor health outcomes? Who suffers the greatest changes in social network structure when marriage ends, and are these changes correlated with health outcomes? How do attachment styles, sex, and relationship history interact to predict physiological reactivity in the laboratory? These are just a few of the questions we can and should be asking. When combined with experimental methods, the field will be able to establish considerable empirical precision about the mechanisms linking marital dissolution and health.

References

Adler, N. E., Boyce, T., Chesney, M. A., Cohen, S., Folkman, S., Kahn, R. L., et al. (1994). Socioeconomic status and health: The challenge of the gradient. *American Psychologist, 49*, 15–24.

Agrawal, A., & Lynskey, M. T. (2009). Correlates of later-onset cannabis use in the National Epidemiological Survey on Alcohol and Related Conditions (NESARC). *Drug and Alcohol Dependence, 105*, 71–75.

Amato, P. R. (2000). The consequences of divorce for adults and children. *Journal of Marriage & the Family, 62*, 1269–1287.

Ben-Shlomo, Y., Smith, G. D., Shipley, M., & Marmot, M. G. (1993). Magnitude and causes of mortality differences between married and unmarried men. *Journal of Epidemiology Community Health, 47*, 200–205.

Berkman, L. F., Glass, T., Brissette, I., & Seeman, T. E. (2000). From social integration to health: Durkheim in the new millennium. *Social Science & Medicine, 51*, 843–857.

Bonanno, G. A. (2004). Loss, trauma, and human resilience: Have we underestimated the human capacity to thrive after extremely aversive events? *American Psychologist, 59*, 20–28.

Bower, J. E., Kemeny, M. E., Taylor, S. E., & Fahey, J. L. (1998). Cognitive processing, discovery of meaning, CD4 decline, and AIDS-related mortality among bereaved HIV-seropositive men. *Journal of Consulting & Clinical Psychology,* 66, 979–986.

Bramlett, M. D., & Mosher, W. D. (2002). *Cohabitation, marriage, divorce, and remarriage in the United States*. Hyattsville, MD: National Center for Health Statistics.

Cheung, Y. B. (1998). Can marital selection explain the differences in health between married and divorced people? From a longitudinal study of a British birth cohort. *Public Health, 112*, 113–117.

Cheung, Y. B. (2000). Marital status and mortality in British women: A longitudinal study. *International Journal of Epidemiology, 29*, 93–99.

Cohen, S., & Herbert, T. B. (1996). Health psychology: Psychological factors and physical disease from the perspective of human psychoneuroimmunology. *Annual Review of Psychology, 47*, 113–142.

Cohen, S., & Janicki-Deverts, D. (2009). Can we improve our physical health by altering our social networks? *Perspectives on Psychological Science, 4*, 375–378.

Cohen, S., & Rodriguez, M. S. (1995). Pathways linking affective disturbances and physical disorders. *Health Psychology, 14*, 374–380.

Dawson, D. A., Grant, B. F., Stinson, F. S., & Chou, P. S. (2006). Maturing out of alcohol dependence: The impact of transitional life events. *Journal of Studies on Alcohol, 67*, 195–203.

Dickerson, S. S., & Kemeny, M. E. (2004). Acute stressors and cortisol responses: A theoretical integration and synthesis of laboratory research. *Psychological Bulletin, 130*, 355–391.

Doi, Y., Minowa, M., Okawa, M., & Uchiyama, M. (2000). Prevalence of sleep disturbance and hypnotic medication use in relation to sociodemographic factors in the general Japanese adult population. *Journal of Epidemiology, 10*, 79–86.

D'Onofrio, B. M., Turkheimer, E. N., Eaves, L. J., Corey, L. A., Berg, K., Solaas, M. H., et al. (2003). The role of the children of twins design in elucidating causal relations between parent characteristics and child outcomes. *Journal of Child Psychology and Psychiatry, 44*, 1130–1144.

Dupre, M. E., Beck, A. N., & Meadows, S. O. (2009). Marital trajectories and mortality among US adults. *American Journal of Epidemiology, 170*, 546–555.

Ebrahim, S., Wannamethee, G., McCallum, A., Walker, M., & Shaper, A. G. (1995). Marital status, change in marital status, and mortality in middle-aged British men. *American Journal of Epidemiology, 142*, 834–842.

Eng, P. M., Kawachi, I., Fitzmaurice, G., & Rimm, E. B. (2005). Effects of marital transitions on changes in dietary and other health behaviours in U.S. male health professionals. *Journal of Epidemiology Community Health, 59*, 56–62.

Ertel, K. A., Glymour, M. M., & Berkman, L. F. (2009). Social networks and health: A life course perspective integrating observational and experimental evidence. *Journal of Social and Personal Relationships, 26*, 73–92.

Farr, W. (1858). The influence of marriage on the mortality of the French people. *Transactions of the National Association for the Promotion of Social Science*, 504–513.

Fu, H., & Goldman, N. (2000). The association between health-related behaviours and the risk of divorce in the USA. *Journal of Biosocial Science, 32*, 63–88.

Goldman, N. (1993). Marriage selection and mortality patterns: Inferences and fallacies. *Demography, 30*, 189–208.

Hajak, G. (2001). Epidemiology of severe insomnia and its consequences in Germany. *European Archives of Psychiatry and Clinical Neuroscience, 251*, 49–56.

Hetherington, E. M., & Kelly, J. (2002). *For better or for worse: Divorce reconsidered.* New York: Norton & Company.

Higgins, J., & Green, S. (Eds.). (2009). *Cochrane handbook for systematic reviews of interventions version 5.0.2 [updated September, 2009]*: The Cochrane Collaboration, 2009. Available from http://www.cochrane-handbook.org

Holmes, T. H., & Rahe, R. H. (1967). The social readjustment rating scale. *Journal of Psychosomatic Research, 11*, 213–218.

House, J. S., Landis, K. R., & Umberson, D. (1988). Social relationships and health. *Science, 241*, 540–545.

Hu, Y. R., & Goldman, N. (1990). Mortality differentials by marital-status: An international comparison. *Demography, 27*, 233–250.

Ikeda, A., Iso, H., Toyoshima, H., Fujino, Y., Mizoue, T., Yoshimura, T., et al. (2007). Marital status and mortality among Japanese men and women: The Japan collaborative cohort study. *BMC Public Health, 7*, 73.

Iwasaki, M., Otani, T., Sunaga, R., Miyazaki, H., Xiao, L., Wang, N. et al. (2002). Social networks and mortality based on the Komo-Ise cohort study in Japan. *International Journal of Epidemiology, 31*, 1208–1218.

Jockin, V., McGue, M., & Lykken, D. T. (1996). Personality and divorce: A genetic analysis. *Journal of Personality and Social Psychology, 71*, 288–299.

Johnson, N. J., Backlund, E., Sorlie, P. D., & Loveless, C. A. (2000). Marital status and mortality: The national longitudinal mortality study. *Annals of Epidemiology, 10*, 224–238.

Joung, I. M., Stronks, K., Van de Mheen, H., & Mackenbach, J. P. (1995). Health behaviours explain part of the differences in self reported health associated with partner/marital status in the Netherlands. *Journal of Epidemiology and Community Health, 49*, 482–488.

Kalmijn, M., & van Groenou, M. B. (2005). Differential effects of divorce on social integration. *Journal of Social and Personal Relationships, 22*, 455–476.

Kemeny, M. E. (2003). The psychobiology of stress. *Current Directions in Psychological Science, 12*, 124–129.

Kiecolt-Glaser, J. K., Fisher, L. D., Ogrocki, P., Stout, J. C., Speicher, C. E., & Glaser, R. (1987). Marital quality, marital disruption, and immune function. *Psychosomatic Medicine, 49*, 13–34.

Kiecolt-Glaser, J. K., Kennedy, S., Malkoff, S., Fisher, L., Speicher, C. E., & Glaser, R. (1988). Marital discord and immunity in males. *Psychosomatic Medicine, 50*, 213–229.

Kiecolt-Glaser, J. K., Marucha, P. T., Malarkey, W., Mercado, A. M., & Glaser, R. (1995). Slowing of wound healing by psychological stress. *Lancet*, 346, 1194–1196.

Kiecolt-Glaser, J. K., McGuire, L., Robles, T. F., & Glaser, R. (2002). Emotions, morbidity, and mortality: New perspectives from psychoneuroimmunology. *Annual Review of Psychology, 53*, 83–107.

Kposowa, A. J. (2000). Marital status and suicide in the national longitudinal mortality study. *Journal of Epidemiology Community Health, 54*, 254–261.

Lee, S., Cho, E., Grodstein, F., Kawachi, I., Hu, F. B., & Colditz, G. A. (2005). Effects of marital transitions on changes in dietary and other health behaviours in US women. *International Journal of Epidemiology, 34*, 69–78.

Lund, R., Christensen, U., Holstein, B. E., Due, P., & Osler, M. (2006). Influence of marital history over two and three generations on early death. A longitudinal study of Danish men born in 1953. *Journal of Epidemiology Community Health, 60*, 496–501.

Malyutina, S., Bobak, M., Simonova, G., Gafarov, V., Nikitin, Y., & Marmot, M. (2004). Education, marital status, and total and cardiovascular mortality in Novosibirsk, Russia: A prospective cohort study. *Annals of Epidemiology, 14*, 244–249.

Mancini, A. D., Bonanno, G. A., & Clark, G. E. (2009). *Stepping off the hedonic treadmill: Individual differences in response to major life events.* Manuscript submitted for publication.

Manzoli, L., Villari, P., Pirone, G. M., & Boccia, A. (2007). Marital status and mortality in the elderly: A systematic review and meta-analysis. *Social Science & Medicine, 64*(1), 77–94.

Mastekaasa, A. (1997). Marital dissolution as a stressor: Some evidence on psychological, physical, and behavioral changes in the pre-separation period. *Journal of Divorce & Remarriage, 26*, 155–183.

Mazur, A., & Michalek, J. (1998). Marriage, divorce, and male testosterone. *Social Forces, 77*, 315–330.

McEwen, B. S. (1998). Protective and damaging effects of stress mediators. *New England Journal of Medicine, 338*, 171–179.

McGue, M., & Lykken, D. T. (1992). Genetic influence on risk of divorce. *Psychological Science, 3*, 368–373.

Mendes De Leon, C. F., Appels, A. D., Otten, F. W., & Schouten, E. G. (1992). Risk of mortality and coronary heart disease by marital status in middle-aged men in the Netherlands. *International Journal of Epidemiology, 21*, 460–466.

Milardo, R. M. (1987). Changes in social networks of women and men following divorce: A review. *Journal of Family Issues, 8*, 78–96.

Miller, G. E., & Blackwell, E. (2006). Turning up the heat: Inflammation as a mechanism linking chronic stress, depression, and heart disease. *Current Directions in Psychological Science, 15*, 269–272.

Miller, G. E., Cohen, S., Rabin, B. S., Skoner, D. P., & Doyle, W. J. (1999). Personality and tonic cardiovascular, neuroendocrine, and immune parameters. *Brain, Behavior, and Immunity, 13*, 109–123.

Miller, T. Q., Smith, T. W., Turner, C. W., Guijarro, M. L., & Hallet, A. J. (1996). Meta-analytic review of research on hostility and physical health. *Psychological Bulletin, 119*, 322–348.

Molloy, G. J., Stamatakis, E., Randall, G., & Hamer, M. (2009). Marital status, gender and cardiovascular mortality: Behavioural, psychological distress and metabolic explanations. *Social Science & Medicine, 69*, 223–228.

Nilsson, P. M., Nilsson, J. A., Ostergren, P. O., & Berglund, G. (2005). Social mobility, marital status, and mortality risk in an adult life course perspective: The Mälmo preventive project. *Scandinavian Journal of Public Health, 33*, 412–423.

Osler, M., McGue, M., Lund, R., & Christensen, K. (2008). Marital status and twins' health and behavior: An analysis of middle-aged Danish twins. *Psychosomatic Medicine, 70*, 482–487.

Peterson, R. R. (1996). A re-evaluation of the economic consequences of divorce. *American Sociological Review, 61*, 528–536.

Powell, L. H., Lovallo, W. R., Matthews, K. A., Meyer, P., Midgley, A. R., Baum, A., et al. (2002). Physiologic markers of chronic stress in premenopausal, middle-aged women. *Psychosomatic Medicine, 64*, 502–509.

Rands, M. (1988). Changes in social networks following marital separation and divorce. In R. M. Milado (Ed.), *Families and social networks* (pp. 127–146). Newbury Park, CA: Sage.

Roberts, B. W., Kuncel, N. R., Shiner, R., Caspi, A., & Goldberg, L. R. (2007). The power of personality: The comparative validity of personality traits, socioeconomic status, and cognitive ability for predicting important life outcomes. *Perspectives on Psychological Science, 2*, 313–345.

Robles, T. F. (2007). Stress, social support, and delayed skin barrier recovery. *Psychosomatic Medicine, 69*, 807–815.

Sapolsky, R. M. (1996). Why stress is bad for your brain. *Science, 273*, 749–750.

Sapolsky, R. M. (1998). *Why zebras don't get ulcers*. New York: W. H. Freeman and Company.

Saudino, K. J., Pedersen, N. L., Lichtenstein, P., McClearn, G. E., & Plomin, R. (1997). Can personality explain genetic influences on life events? *Journal of Personality and Social Psychology, 72*, 196–206.

Sbarra, D. A., & Emery, R. E. (2005). Co-parenting conflict, nonacceptance, and depression among divorced adults: Results from a 12-year follow-up study of child custody mediation using multiple imputation. *American Journal of Orthopsychiatry, 75*, 63–75.

Sbarra, D. A., & Hazan, C. (2008). Co-regulation, dysregulation, and self-regulation: An integrative analysis and empirical agenda for understanding attachment, separation, loss, and recovery. *Personality & Social Psychology Review, 12*, 141–167.

Sbarra, D. A., Law, R. W., Lee, L. A., & Mason, A. E. (2009). Marital dissolution and blood pressure reactivity: Evidence for the specificity of emotional intrusion-hyperarousal and task-rated emotional difficulty. *Psychosomatic Medicine, 71*, 532–540.

Sbarra, D. A., & Nietert, P. J. (2009). Divorce and death: Forty years of the Charleston heart study. *Psychological Science, 20*, 107–113.

Scarr, S., & McCartney, K. (1983). How people make their own environments: A theory of genotype environment effects. *Child Development, 54*, 424–435.

Segerstrom, S. C., & Miller, G. E. (2004). Psychological stress and the human immune system: A meta-analytic study of 30 years of inquiry. *Psychological Bulletin, 130*, 601–630.

Shear, K., & Shair, H. (2005). Attachment, loss, and complicated grief. *Developmental Psychobiology, 47*, 253–267.

Stroebe, M., Schut, H., & Stroebe, W. (2007). Health outcomes of bereavement. *Lancet, 370*, 1960–1973.

Tejada-Vera, B., & Sutton, B. (2009). Births, marriages, divorces, and deaths: Provisional data for July 2008. *National Vital Statistics Reports*, 57, 13. Hyattsville, MD: National Center for Health Statistics.

Tomaka, J., Blascovich, J., Kelsey, R. M., & Leitten, C. L. (1993). Subjective, physiological, and behavioral effects of threat and challenge appraisal. *Journal of Personality & Social Psychology, 65*, 248–260.

Tucker, J. S., Friedman, H. S., Wingard, D. L., & Schwartz, J. E. (1996). Marital history at midlife as a predictor of longevity: Alternative explanations to the protective effect of marriage. *Health Psychology, 15*, 94–101.

Umberson, D. (1987). Family status and health behaviors: Social control as a dimension of social integration. *Journal of Health and Social Behavior, 28*, 306–319.

Umberson, D. (1992). Gender, marital status and the social control of health behavior. *Social Science & Medicine, 34*, 907–917.

Uunk, W. (2004). The economic consequences of divorce for women in the European Union: The impact of welfare state arrangements. *European Journal of Population/Revue Européenne de Démographie, 20*, 251–285.

Wade, T. J., & Pevalin, D. J. (2004). Marital transitions and mental health. *Journal of Health & Social Behavior, 45*, 155–170.

Waite, L. J., & Gallagher, M. (2000). *The case for marriage: Why married people are happier; healthier; and better off financially*. New York: Broadway Books.

Weitzman, L. J. (1985). *The divorce revolution: The unexpected social and economic consequences for women and children in America*. New York: Free Press, Collier Macmillan.

Whisman, M. A., Tolejko, N., & Chatav, Y. (2007). Social consequences of personality disorders: Probability and timing of marriage and probability of marital disruption. *Journal of Personality Disorders, 21*, 690–695.

Williams, K., & Umberson, D. (2004). Marital status, marital transitions, and health: A gendered life course perspective. *Journal of Health and Social Behavior, 45*, 81–98.

Wilmoth, J., & Koso, G. (2002). Does marital history matter? Marital status and wealth outcomes among preretirement adults. *Journal of Marriage and the Family, 64*, 254–268.

Zagorsky, J. L. (2005). Marriage and divorce's impact on wealth. *Journal of Sociology-Australian Sociological Association, 41*, 406–424.

CHAPTER 12

The Future of Relationship Science

JOHN G. HOLMES

University of Waterloo

The field of relationship research has grown exponentially in recent decades. Progress has been especially impressive in terms of the development of sophisticated new theoretical frameworks and analytical techniques that are capable of reflecting the complexity of relationships. In this chapter, I will first provide an overview of the important new directions for research evident in the collection of presentations (and chapters) for the "Science of the Couple" workshop and then provide a discussion of four trends in these domains that I personally believe are central to the future of the science.

The Science of the Couple Workshop

The first two editors, Lorne Campbell and Jennifer LaGuardia, organized the workshop to highlight both well-established research trends and important new directions of research. They also put an emphasis, within these categories, on novel and dynamic research techniques, methods that often allow researchers to reach for new and unexplored horizons.

Their first category highlighted personality approaches. The personality perspective and its developmental roots have been a major, if not the major, focus for relationship researchers. The contributions of attachment theory have been enormous, and the breadth and depth of its influence are well reflected in the two chapters on personality by some of the foremost theorists. First, Shaver and Mikulincer, major pioneers in the field, provide a roadmap of the development of ideas from Bowlby onward, illustrating the incredible richness of this theoretical foundation. They provide one of the first thorough discussions of anxious attachment in adult romantic relationships, focusing on the unique motivational patterns of individuals high in attachment anxiety. Tran and Simpson then provide a cutting-edge analysis of how a person–situation perspective is useful in understanding attachment. They suggest that certain situations, such as a conflict with a partner, raise the issue of whether a partner can be counted on to be responsive to one's needs. Given that this issue is central to those with anxious attachment, such situations appear to strongly activate the defensive systems of such individuals. Tran and Simpson also further theoretical development by integrating ideas from interdependence theory with attachment, showing how different levels of commitment interact with attachment styles in predicting behavior.

The next grouping of researchers reflects a more social psychological approach, with a particular focus on processes that promote healthy relationship functioning. Fitzsimons and Finkel discuss relational goals and the fascinating dyadic issue of how one person's efforts or ability to complete a goal might influence the other's efforts and goal persistence. They propose the novel idea that in one sense, partners become "substitutable" in their goal pursuits, with one person's efforts in goal pursuit actually undermining the other's motivations. Murray and Pinkus, like Fitzsimons and Finkel, emphasize recent social cognitive principles and how they function in relationships. They illustrate how automatic goal activation can be overturned by more deliberative processes. Specifically, they illustrate how these processes depend on people's opportunities and motivation to establish their preferred chronic agenda regarding motivations for closeness or distance in their relationships. Finally, Gable analyzes specific processes in social interaction related to intimacy and closeness in relationships. She puts particular emphasis on the positive consequences for intimacy of "capitalizing" with a partner—that is, responsively sharing and enjoying a partner's successes and positive experiences. Perhaps surprisingly, she illustrates how these processes may be even more important for the health of a relationship than providing support for dealing with a partner's negative and stressful experiences.

The next category of chapters reflects very recent and productive trends toward interdisciplinary models of relationship functioning. Indeed, all three chapters in this category offer quite unique contributions that depend on the most recent advances in new methodologies. First, the chapter by Loving and Wright applies new insights from psychoneuroimmunology and the physiology of passionate love to understanding relational "stress." They discuss research that considers the context in which cortisol responses are observed to best interpret the meaning of any observed changes during acute and chronic social settings. Loving and Wright note recent work that suggests that increased cortisol does not represent a stress response, per se, insofar as any increase necessarily contributes to poor health outcomes. Next, the chapter by Beckes and Coan focuses on the neuroscience and biology of emotion regulation. Their social baseline theory (SBT) proposes that the default mode of human affect regulation is through social proximity and interaction, effects that are mediated through subcortical neural circuits including the amygdala, the nucleus accumbens, and the ventral tegmentum. Because self-regulatory efforts mediated through the prefrontal cortex are metabolically costly, SBT predicts that social emotion regulation strategies drive many manifestations of social proximity to restore equilibrium in the brain. Finally, the Sbarra chapter reviews three primary pathways that may explain the established connection between marital separation/divorce and poor physical health outcomes. He presents a meta-analytic review of over 20 large-scale epidemiological studies that document the magnitude of the divorce–health effect in a variety of different outcome domains. Sbarra then considers the particular ways the psychological stress of divorce may contribute to poor health. These three chapters illustrate the important advances in thinking about relationships made possible by advances in the biological side of the science.

In the final category, the focus is on the critical topic of well-being and health that I suspect will become a major theme in future years. The Lydon and Linardatos chapter adopts the working assumption that, given the strong link between close relationships and health and the suggestion that relatedness is an intrinsic psychological need, it is not surprising that people often identify with their intimate relationships. Lydon and Linardatos present a theoretical framework for understanding how relationship identities develop and what consequence they have for personal and relationship well-being. They describe a series of studies on relationship identification and its effects on relationship satisfaction and survival as well as the possible adverse effects of lingering identification post-breakup. The La Guardia chapter breaks new ground by studying how people cope with serious life crises. She considers couples' adjustment to life following a cardiac crisis and found that, the greater the emotional impact of the heart event for the patient, the more their spouse also blocked their own negative feelings about the event, thereby colluding in "not looking" at the emotional challenges brought on by the cardiac crisis. The study showed that when patients willingly process their negative emotions rather than block them, they experience greater psychological health as well as greater closeness and satisfaction with their partner. These two chapters present evocative evidence on the link between the quality of relationship processes and health that opens up important new territory for relationship researchers.

I was left feeling almost overwhelmed by the breadth and depth of the work presented in the workshop. The future of relationship science looks bright indeed. To cut across this complexity, I next want to focus on four trends that I see in recent developments in the field that cut across the various categories previously described, trends that I believe will be central to the future of the science.

Four Significant Trends

I first discuss recent developments in analytical and statistical models in two important domains. New modeling techniques have been designed to deal with interdependence between partners' data and also to analyze results from within-person designs (e.g., daily diaries, longitudinal studies) that measure interactions between partners over time. Second, I describe recent efforts to categorize the elements of social situations. This theoretical work by interdependence theorists provides the basis for person–situation models that link social processes with personality (see Reis & Holmes, in press) and for specifying situational cues that shape social cognitive processes such as goal activation. Third, I suggest that it is crucial for relationship researchers to incorporate into their frameworks dual process models that are prominent in the area of social cognition. I describe research on attachment and on risk regulation that has made serious headway in studying these automatic processes. Finally, I make a plea for more theoretical integration in relationship science. I suggest that there are core concepts, such as expected responsiveness, that are common to various frameworks. The conceptual commonalities make clear that the field is making considerable progress in developing a broad, overarching theoretical framework for understanding interpersonal processes.

Developments in Analytical Techniques

A unique set of challenges faces researchers who study social interactions or relationship processes. Data from one member of a dyad are not independent from that of the other person, requiring estimates of the degree of interdependence to correct for mutual influence or shared third variables. This problem, coupled with the fact that many studies on relationships are essentially of a complex correlational nature, means that relationship researchers often face a far more daunting task in terms of statistical analyses than other areas. To the considerable credit of our colleagues, there have probably been more serious advances in statistical methods emanating from relationship research than from any other area in psychology.

Kelley (1979) first outlined the logical issues related to interpersonal interdependence, but recently our colleagues have figured out how to account for dyadic sources of variance (see Campbell in this volume for a review). For example, Kenny and colleagues' actor-partner interdependence model (APIM; Kenny, Kashy, & Cook, 2006) makes the point that it is not just a need to adjust estimates for patterns of influence that is at issue, but rather we need to directly estimate cross-person influences. That is, a variable related to an actor may have effects not only on the actor but also on the partner and vice versa. So for example, an actor's high level of commitment may result in the actor resisting attractive alternatives by devaluing them (see Rusbult & Van Lange, 2003) but also affect a *partner's* confidence or anxiety about the state of the relationship, through the actor's commitment-related behavior and reactions (Simpson, Ickes, & Blackstone, 1995). Such interpersonal effects lie at the very heart of the discipline and can now be clearly identified. This sophisticated new modeling therefore opens up the possibility of studying a wide variety of questions that could not previously be accomplished.

My favorite example of the ability of these new analytical tools to untangle difficult issues involves research by Cook (2000) on attachment orientations. Cook's research was based on an early version of Kenny's reasoning on partitioning sources of interpersonal influence, the social relations model (SRM; Kenny, 1994). As background to Cook's work, Holmes and Cameron (2005) argued that a critical, largely unresolved issue is the level of generality in attachment representations of self and others. Collins and Read (1994) pointed out that attachment theorists have generally been rather vague about how abstract models are organized in cognitive networks that might also contain representations of *specific* relationships. Indeed, most theorists have simply portrayed people as having a single set of models and a unique attachment style.

Collins and Read (1994) proposed a hierarchical framework with different levels of abstraction, ranging from general working models to generalizations about particular significant others to differentiated representations of specific others in specific roles or situations. They described the general model as the "default option," a global style that develops from early experience with caregivers. This is in contrast to the much more typical view that general models exert top-down, assimilative pressures on perceptions in specific new attachment relationships (e.g., Hazan & Shaver, 1987). Attachment theorists have almost always treated attachment orientations as individual differences or traits, that is, as general models. In contrast, Murray and her colleagues have focused on the degree of security or insecurity experienced in a specific close relationship as the critical variable (Murray, Holmes, & Collins, 2006).

The research by Cook (2000) extended Kenny's analytical thinking to explore these important distinctions about the generality of people's expectations. He studied individuals' ratings of attachment security in over 200 families with two teenage children. Because each person could rate his or her sense of security with each of the other family members, this multipartner assessment allowed variance to be partitioned by the source of felt security or insecurity. Importantly, Cook found consistency in attachment security across relationships, indicating that actors' *general* working models indeed do have an influence on expectations about specific relationships. If a mother felt insecure about the affections of one member of the family, she tended to report the same concerns with other family members.

He also found fascinating and novel evidence for "partner" effects, meaning that certain people tended to *induce* insecurity (or security) in most others. Interestingly, this finding supports the view that interpersonal expectations are not simply "all-in-the-head" constructions but also reflect objective (or at least intersubjective) aspects of the social environment (i.e., the qualities of actual partners).

Cook (2000) also found that a person's reported attachment security had a component uniquely tied to a particular partner, a relationship-specific or dyadic effect. These "interpersonal chemistry" effects were just as strong as the individual difference effects. Everyone (from the different theoretical vantage points) gets to eat cake. That is, both general working models and specific beliefs about partners are important to consider.

Other developments have extended structural equation modeling (SEM) in interesting new directions, again opening up new domains for exploration (Griffin & Gonzalez, 2003). SEM has been particularly important for analyzing longitudinal data and effects over time. However, perhaps the most critical advance has been the application and refinement of hierarchical linear modeling (HLM). HLM is essentially a form of within-person regression that permits researchers to analyze people's actions over time while at the same time examining chronic or stable moderating variables such as attachment or self-esteem. Such designs can be used to describe trajectories of change over longer periods of time, such as a 5-year period in marriages (e.g., McNulty & Karney, 2004). They can also open up the rich world of people's everyday experiences through asking couples to complete daily diaries describing their thoughts, feelings, and behavioral reactions.

Reis and Judd (2000) argued that within-person designs are extremely powerful tools in the sense that the person is essentially serving as his or her own control group in establishing baselines, controlling for all sorts of extraneous sources of variance and holding chronic traits constant. For instance, Murray, Holmes, Aloni, Pinkus, Derrick, and Leder (2009) used a daily diary methodology to track changes in people's feelings of inferiority in their relationships. They demonstrated that such feelings were associated with people's efforts to increase their instrumental contributions, presumably to compensate for diminished sense of "value." These instrumental contributions were shown to in fact be related to increases in partner satisfaction on a day-by-day basis. The precision of the within-person design meant that even subtle variations in day-to-day well-being could be predicted.

Further, the within-person design is a wonderful technique for examining the dynamics of person–situation models (Mischel & Shoda, 1995). In person–situation models, the nature of situational features encountered each day can be indexed and linked to people's perceptions and personality traits. For instance, I will later discuss research by Overall and Sibley (2009) that links an insecure attachment style to reductions in closeness on days when the person's outcomes are heavily dependent on a partner's actions and goodwill. Murray, Griffin, Rose, and Bellavia (2003) demonstrated that low-self-esteem individuals in particular reacted to feelings of rejection on a particular day by distancing themselves from their partner. Finally, Campbell, Simpson, Boldry, and Rubin (2010) showed that individuals who were uncertain about trusting in their partners' feelings for them were more "reactive" to everyday events in the relationship, amplifying the meaning of both positive and negative experiences. While hierarchical modeling has many uses, its capability of modeling person–situation links is perhaps its most critical contribution to sophisticated theorizing in the field of relationships.

Features of Situations: Person–Situation Models and Social Cognition

Research on relationships has traditionally focused on both personality factors and social psychological situational causes as separate predictors of relational outcomes but only recently has focused on how these elements might function together. A conceptualization that incorporates such a person-by-situation logic has been called an interactionist perspective (Cantor & Kihlstrom, 1987; Endler & Hunt, 1969). This perspective has received a strong boost from the development of Mischel and Shoda's (1995) "meta"-theoretical model that linked persons and situations through the cognitive-affective personality system (CAPS). The CAPS framework has had a strong influence on the field of personality psychology in recent years, in part because it provided a more general social-cognitive interpretation of the meaning of "personality."

Mischel and Shoda (1995) present impressive evidence that an individual's "behavioral signature" is typically quite stable over time if behavior is examined within the context of *specific* situations. A person's signature is composed of "if–then" patterns of situation-behavior associations (see Figure 12.1). The CAPS model contends that specific features of situations activate subsets of cognitive mediating units, which in turn generate responses to the different situations. That is, individuals are seen to have

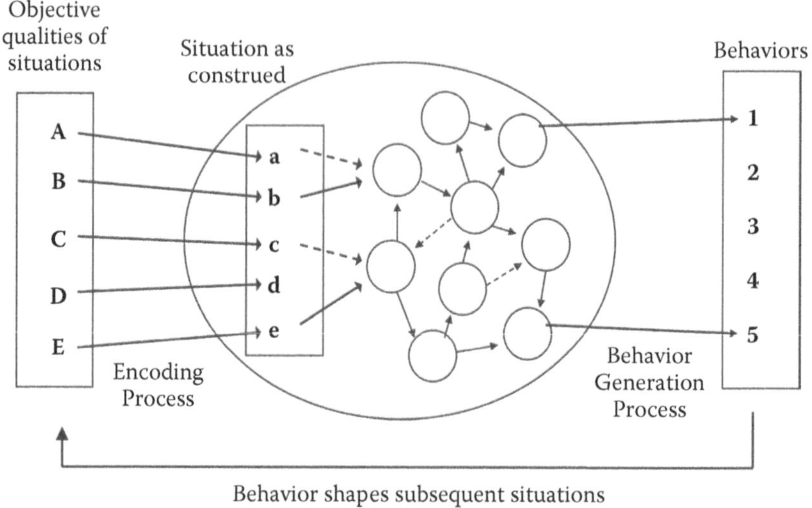

Figure 12.1 Cognitive–affective personality system (CAPS).

a distinctive behavioral signature or style of adapting to features of their social environment. In this regard, Mischel and Shoda suggest that situations need to be considered in abstract terms, redefining them "to capture their basic psychological features, so that behavior can be predicted across a broad range of contexts that contain the same features" (p. 248).

I strongly concur with this principle that features of situations need to be categorized a priori in terms of their psychologically critical, "active" ingredients (and indeed I developed this as a major theme in a paper on the structure of interpersonal cognition; Holmes, 2002). If instead we focus on concrete or nominal details of situations, we easily become lost in the minutiae of everyday life and lose predictive power. However, Mischel and Shoda's (1995) definition of the situation, the *if* in their if–then model, has not been particularly abstract but instead has focused largely on another person's behavior as the context for the actor's behavior. For example, one "situation" in their famous summer camp study was, "Adult warned the child." This perspective is certainly not unreasonable given that focusing on what people see and hear in the social world is the most common usage of the term *situation* in social psychology (Ross & Nisbett, 1991). However, it is difficult to know what an abstract depiction of this situation would be. For instance, is the counselor giving an authoritative command with possible consequences, criticizing the child's behavior, or instead trying to coordinate swimming activities? In this situation, we might predict very different responses from the child depending on our theoretical orientation.

The problem is that we need a "theory of situations" if we are to create a model that starts with specifying abstract features of external situations and then predicts a priori which elements in the cognitive-affective system will be activated to adapt to or cope with the opportunities or challenges the social situation *affords* by its structure (see Reis & Holmes, in press). The activated elements can then be related to goal processes that predict the behavior generation process. I turn to recent developments in interdependence theory to explore how a top-down theory of situations might provide a theoretical classification of situations and an analysis of the particular cognitive and motivational processes functionally related to dealing with the specific problems each situation entails.

Interdependence Theory

My description of interdependence theory (IT) is adapted from *An Atlas of Interpersonal Situations* by Kelley, Holmes, Kerr, Reis, Rusbult, and Van Lange (2003). This book analyzes 20 of the most prototypical social situations in detail and presents propositions linking each to particular cognitive and goal processes as well as to individual differences in personality. It expands on the long intellectual tradition of social exchange analysis first presented by Thibaut and Kelley (1959) and later elaborated by Kelley and Thibaut (1978). Some implications of this general theory for social cognitive and personality processes were developed by Holmes (2002).

Generally speaking, IT expands the formula proposed by Lewin (1946) that behavior is a function of the person and the environment. In the context of social interaction, the behavioral interaction (I) that occurs between persons A and B is thought to be a function of both persons' respective goal tendencies in relation to each other *in the particular situation of interdependence* (S) in which the interaction occurs.

By the phrase *particular situation of interdependence*, IT refers to the ways two persons depend on and influence each other with respect to their potential outcomes from an interaction (hence the term interdependence). The theory attempts to identify the kinds of interpersonal dispositions of persons A and B—their attitudes, motives, goals—that are functionally relevant to each particular type of situation. The type of situation S, together with the relevant dispositions of persons A and B, determine the interaction, I (in symbols, the SABI elements; Holmes, 2000). As this model implies, interdependence theory adopts a Person *x* Situation interactionist approach with a strong social-psychological focus on the nature of situations, such that each paradigmatic situation is viewed as presenting the two persons with a unique set of problems and opportunities.

The study of situations revealed six basic dimensions, to be discussed herein, that can be used to categorize any particular social problem. To illustrate the fundamental importance of these dimensions to social psychology, Reis (2008) and Reis and Holmes (in press) showed in detail how these dimensions can be used to categorize a wide variety of research topics in social psychology. These examples suggest that social psychology has long understood (though often more intuitively than explicitly) that these dimensions are important and that in everyday life people discriminate among them naturally and fluidly.

Four of the six dimensions of outcome interdependence can be logically deduced from the structure of decisions in social dilemmas:

- The extent to which an individual's outcomes depend on the actions of the other person
- The extent to which individuals have mutual power over each other's outcomes or whether power is asymmetric and one person is more dependent on the other's goodwill
- The extent to which one person's outcomes correspond or conflict with the other's
- The extent to which partners must coordinate their activities to produce satisfactory outcomes from shared goals or whether the goals are only partly shared and each one's actions in an exchange partly determine the other's outcomes

Interdependence theory also considers two additional elements of situations:

- Temporal structure: Whether the situation involves interaction over the longer term
- Information certainty: Whether partners have the information needed to make good decisions or whether substantial uncertainty exists about the future

The dimensions are logically related to understanding the dynamics of personality, given that each dimension of situations affords the expression of particular motives and functional adaptations. Holmes and Cameron (2005) presented a meta-analytic framework linking the six dimensions to the particular motives and goals that they afford. By doing so, they then speculate about individual differences on the Big Five and attachment avoidance that would represent comparable chronic motivational agendas (see Table 12.1). In a sense, the synergy between situations and personality is most apparent when theorists explicitly consider in this way the relevance of personality features to the particular affordances of situations.

This IT framework fits particularly well with Mischel and Shoda's (1995) CAPS model linking situations to personality. Indeed, according to Mischel and Morf (2003) many of the most compelling examples of CAPS thinking involve research into interpersonal relations, including studies of rejection sensitivity (Downey, Freitas, Michaelis, & Khouri, 1998) and narcissism (Morf & Rhodevalt, 2001).

Table 12.1 Dimensions of Situations and Interpersonal Dispositions

Dimension of Situation	Function of Rule	Interpersonal Disposition
Degree of interdependence	Increase or decrease dependence on partner	Avoidance of interdependence/
Mutuality of interdependence		Comfort with dependence
Correspondence of outcomes	Promote prosocial or self-interested goals	Cooperative/competitive
		Responsive/unresponsive
	Expectations about partner's goals	Anxiety about responsiveness/
		Confidence or trust
Basis of control	Control through exchange (promise/threat) or coordination (initiative/follow)	Dominant/submissive
		Assertive/passive
Temporal structure	Promote immediate or distant goal striving	Dependable/unreliable
		Loyal/uncommitted
Degree of uncertainty	Cope with incomplete information or uncertain future	Need for certainty/openness
		Optimism/pessimism

Further, because the Atlas dimensions can be used to categorize the features of experimental manipulations, the framework encourages experimental approaches to the study of personality, especially those focusing on how situational activation and affordance reveal latent properties or vulnerabilities in personality structure (diathesis). For instance, certain psychological themes characterize situations that involve a conflict of interest or situations involving high dependence on a partner. In both cases, the issue of a partner's goodwill and caring becomes paramount because the other's self-centered or prosocial motivations will determine a person's own outcomes. Because these two dimensions of situations activate concerns about how others value the self, anxious attachment and low self-esteem become central personality considerations. In both cases, a major dynamic of these personality dispositions is uncertainty or insecurity about being valued.

Simpson, Rholes, and Philips (1996) relied on this logic in the design of a very creative study on attachment styles. They asked established dating couples to discuss either a major problem or conflict or a minor one. In the minor conflict situation, secure and insecure individuals could not be distinguished by their behavior in the interaction. However, in the major conflict of interest condition, the "true colors" of people with an anxious attachment were revealed (Figure 12.2). Anxious individuals assumed the worst about their partner's motivations and acted in an angry, untrusting way, whereas secure individuals assumed the best and reacted even more positively than in the minor conflict situation, presumably because the conflict strongly activated their beliefs about their partner's caring.

In the same vein, Murray, Holmes, and Collins (2006) reported a series of studies on "risk regulation" where they experimentally manipulate situations that amplify concerns about the risk of rejection or nonresponsiveness by a partner. For instance, in one case the manipulation involved

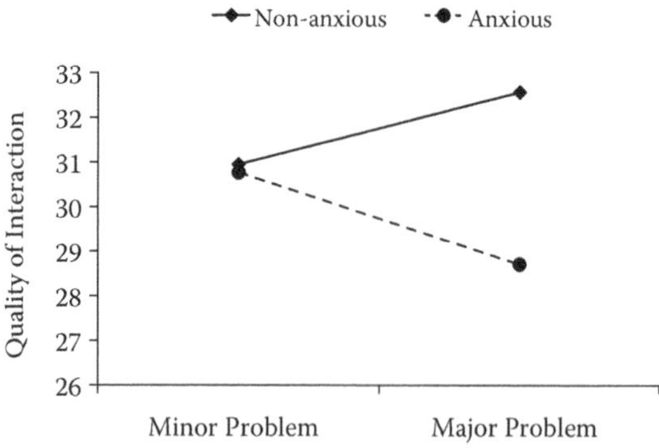

Figure 12.2 The Simpson, Rholes, and Phillips (1996) situation–person interaction pattern.

persuading people that their partner had "hidden complaints" about them. In another, their partner took a great deal of time completing a question supposedly about their most egregious fault (because they actually were asked to describe objects in their living room). Parallel to the findings of Simpson and colleagues, low-self-esteem people in the risk condition, who were uncertain about how they were valued, assumed a stronger likelihood of rejection and then regulated the degree of risk by distancing themselves from their partner, both psychologically and behaviorally. High-self-esteem people actually reported feeling closer to their partner when the risk was greater.

Similar dynamics were observed when concerns of rejection were measured on a daily basis in a diary study of dating couples (Murray, Bellavia, et al., 2003). Situational cues of potential rejection, such as a conflict of interest, arguments, or even a partner's bad mood, resulted in low-self-esteem individuals feeling they were not valued, and, consequently, they distanced themselves from their partner and behaved badly toward their partner (according to both the person and the partner). High-self-esteem individuals were reminded by these risky situations just how valued and loved they felt, and they behaved all the more positively and constructively in the face of risk. Overall and Sibley (2009) similarly conducted daily diary studies with anxiously attached individuals within family, friends, or romantic relationships. They asked people to identify days on which they were particularly *dependent* on the other person for their outcomes. The authors found the same pattern across all three samples—insecure, anxious individuals behaved in a more distant manner on dependent days and felt less close to the other person.

The point of these examples is to impress the value of considering the types of concerns and issues as well as the goal pursuits that each social situation affords. This same logic obviously fits well with recent advances in social cognition that are concerned with the particular cognitions that are activated by social cues, as suggested in the Mischel and Shoda (2005) model (Figure 12.1). It is a particularly good fit with Bargh's (2007) auto-motive model that focuses on goal activation in response to situated cues. Indeed, goal constructs are a powerful tool for integrating notions of the situation with the "person." For instance, Gable (2005) showed how individual differences in social approach and avoidance goal orientations color reactions to and interpretations of positive and negative aspects of social experiences, respectively. Fitzsimons and Finkel (2010; also see Fitzsimons & Finkel, this volume) summarize research on self-regulation and goal activation in relationships, and Fitzsimons and Shah (2008) showed how the success of a social episode is dependent on a partner being categorized as someone who is instrumental in facilitating a person's own situationally activated goals. When goals to pursue one's academic work are activated, for instance, one's party animal friend is avoided. Research linking goal activation and situational features is in its infancy but strikes me as a very promising new direction.

Dual-Process Models of Social Cognition

I believe that research on relationships can benefit greatly from using current new developments in social cognition (and vice versa). Baldwin's (1992) interpersonal schema model, for instance, took ideas about cognitive schema in very creative new directions. He proposed that relational knowledge is represented by a schema for the self in relation to another, a schema for the other in relation to the self, and a relational script specifying if–then rules for expected social interactions. The result was a theory well informed about cognitive processes but also one that extended social cognitive ideas in novel ways.

Recent advances in social cognition have been heavily influenced by dual-process models that consider both people's impulsive reactions (efficient automatic procedural rules) and more deliberative ones (requiring more executive control; see Gawronski & Bodenhausen, 2006; Olson & Fazio, 2009; Strack & Deutsch, 2004; Wilson, Lindsey, & Schooler, 2000). Ongoing and complex problems have automatic and effortless solutions (Bargh, 2007; Dijksterhuis & Nordgren, 2006). The "efficiency" criterion stipulates that associative if–then rules are *implicit* procedural features of relationship representations (Baldwin, 1992; Holmes & Murray, 2007). By implicit, we mean that these rules can operate without conscious mediation (Bargh; Dijksterhuis & Nordgren).

Recent research in interpersonal relations has explored such automatic associative links. Situations that prime the if elicit the propensity to engage in the then without any conscious intent on a person's part. For instance, cues for a stressor outside one's relationship automatically activate connection goals, whereas cues of internal relationship strife automatically activate self-protect or avoidance goals (Murray, Derrick, Leder, & Holmes, 2008). Similarly, cues that signal inferiority result in the activation of goals to contribute more to a relationship (Murray et al., 2009).

There is evidence that automatic attitudes can elicit corresponding behaviors even when contradictory explicit sentiments are accessible in memory (Wilson et al., 2000), what Bargh and Huang (2009) have labeled a "selfish goal." For instance, attachment theorists have demonstrated that unconsciously primed thoughts of attachment security heighten prosocial behavior even in the face of interpersonal risk. People exhibit less outgroup derogation (Mikulincer & Shaver, 2001), increased empathy for others (Mikulincer, Gillath, Halevy, Avihou, Avidan & Eshkoli, 2001), and increased desire to seek support from others in dealing with a personal crisis (Pierce & Lydon, 1998).

This logic implies that positive automatic evaluative responses to one's partner might compel subtle behavioral expressions of commitment, despite considered reasons to be cautious. That is, it is possible that more positive automatic attitudes toward the partner have the power to neutralize deliberative, more conscious expectations about a partner's potential rejection (see Murray & Pinkus, this volume). This might occur because such automatic evaluations better capture the actual structure of rewards (and punishments) that have motivated approach (or avoidance) behaviors in the relationship in the past (Kelley, 1979). In the model presented by Murray, Holmes and Pinkus (2010; also see Murray & Pinkus, this volume) automatic attitudes toward the partner essentially reflected the associative residue of the interdependent mind's relationship "know how" (see Bargh & Huang, 2009).

What is implied by relationship "know how"? Murray and Holmes (2009) attributed a system of inter-connected "if–then" rules to the interdependent mind. These rules *automatically* function to coordinate mutually responsive interactions by linking the risk properties of situations to correspondent interpersonal goals and behavioral strategies for goal pursuit. Imagine that Sally has a difficult week ahead and needs to ask a good-humored Harry to sacrifice a few hours of his free time to spend on childcare. This low-risk situation strengthens her goal to connect and activates behavioral strategies for escalating her dependence on Harry (e.g., asking for his help) and justifies any costs incurred through her greater dependence (e.g., seeing him as all the more thoughtful even though he forgot her dry-cleaning). Imagine instead that Sally needs to ask Harry to sacrifice several days of work time to spend on childcare. This high-risk situation strengthens her goal to self-protect and activates behavioral strategies for withholding her dependence on Harry (e.g., asking the grandparents for help instead of him).

Building on this logic, Murray et al. (2010) hypothesized that repeated experience in more or less risky relationship situations *condition* evaluative associations to the partner by controlling "if–then" habits of thought. In a longitudinal study of newlyweds, these researchers indexed situation-exposure and "if–then" habits of thought in the initial months of marriage through daily diary reports. Then they measured automatic evaluations toward the partner (through the Implicit Associations Test) after 4 years of marriage. The results revealed that the more high-risk situations Sally encountered early on (e.g., Harry doing what he wanted to do over what she wanted to do), the less positive her automatic attitude toward Harry after 4 years. Similarly, the more often Sally reacted to feeling rejected in such situations by distancing herself from Harry (i.e., a self-protective "if–then" rule habit), the less positive her later automatic attitude toward Harry. However, such experiences did not change Sally's explicit beliefs about Harry's responsiveness at all (nor did such explicit beliefs condition her later automatic attitudes).

These contrasting findings suggest that automatic attitudes toward the partner might neutralize deliberative, conscious rejection concerns because such associations better diagnose the concrete rewards of interaction (Murray, et al., 2009). In contrast, Sally's deliberative expectations of Harry's rejection may be quite out of touch with the actual incentives that motivate her behavior (Murray et al., 2000; 2001).

The other side of the dual process model involves flexible adjustment—that is, cognitive corrections to better suit chronic agendas and motivations. Models of attitudes assume that people act on automatic impulses unless they have the motivation *and* opportunity to override such inclinations (Olson & Fazio, 2009). Flexibility implies that an automatic urge to think, feel, or behave in a particular way is less likely to translate into correspondent action if people are motivated *and* able to correct it. In the Murray and Holmes (2009) analysis, *trust* determines how often the automatic propensity to think or behave in a particular way translates into *correspondent* action in specific situations. It does so because trust provides the motivation to correct impulses compelled by "if–then" implementation rules that run counter to a person's more chronic goal orientation toward the relationship (Murray, Aloni, et al., 2009; Murray, Holmes, et al., 2009). Sufficient executive or conscious control over behavior supplies the opportunity to correct (Gilbert & Malone, 1995; Muraven & Baumeister, 2000).

High levels of trust in the partner's responsiveness foster the chronic pursuit of connection in relationships (Mikulincer & Shaver, 2003; Murray et al., 2006). People who are more trusting can more readily afford to set aside self-protection goals in mixed-motive situations because the partner's nonresponsiveness is not that likely or that hurtful (Murray et al., 2003). Low levels of trust in the partner's responsiveness instead foster the *chronic* pursuit of self-protection in relationships. People who are less trusting can less readily afford to pursue connectedness goals in mixed-motive situations because the partner's nonresponsiveness is more likely and more hurtful (Holmes & Rempel, 1989; Murray et al., 2006). Because trust shapes chronic goal pursuit, it provides the motivation to correct the behavioral impulse to pursue contradictory state *goals*.

For instance, being less trusting might motivate a husband to curb his automatic impulse to value a partner more when she interferes with his goal pursuits (Murray et al., 2009), because it risks additional closeness. Similarly, uncertainty about trusting a partner results in people correcting their automatic inclination to approach a partner after experiencing a stressor, whereas trusting partners enact that goal (Murray et al., 2008). Finally, people who momentarily feel inferior compared to their partner tend to contribute more to the relationship to compensate for their poor intrinsic value, a goal which low-trust people actually execute. Trusting partners overturn the goal on the grounds that such costly behavior is an unnecessary waste of resources (Murray, Aloni, Holmes, Derrick, Stinson, & Leder, 2009).

In summary, efforts to explore the application of current ideas on social cognition to models of relationships have already proven valuable, and I suspect there is room for many more productive avenues.

Theoretical Integration

> Psychologists tend to treat other people's theories like toothbrushes; no self-respecting individual wants to use anyone else's.
>
> **Mischel (2008)**

Mischel's amusing but insightful comment could certainly be applied to relationship science in my opinion. There is very little integration among the four or five major frameworks. Indeed, scientists centered in one perspective seldom even cite other theories or evidence from different vantage points. This occurs despite some degree of agreement in the core construct in theories of interpersonal relations—perceived or expected responsiveness (Bowlby, 1982; Holmes & Cameron, 2005; Reis, Clark, & Holmes, 2004). Bowlby's most fundamental principle was that feelings of security in the expected responsiveness and caring of a significant other are central to understanding the cognitive and behavioral adjustments people make in their relationships.

Cameron and Holmes (2009) argued that the construct of expected or perceived responsiveness (PR) is evident in a variety of theoretical approaches, though often in different guises. In this research, we created a new measure of PR to make comparisons across theories. The PR measure had three components that all correlated about 0.80 with each other: expected responsiveness, felt security, and perceived regard. These constructs are thus essentially equivalent psychological states but simply expressed as expected partner behavior, own emotions, and cognitions or beliefs about partner's views, respectively. What potentially corresponding concepts from other theories were examined?

In attachment theory, PR or "felt security" has two components according to Bowlby—model of self and model of other. (There is no direct measure of secure feelings in the literature.) In current measurement terms, this translates into *not* being anxious about caregivers' caring for the self and *not* being avoidant because of discomfort about their goodwill (see Fraley, Waller, & Brennan, 2000). The additive and interactive aspects of the two dimensions supposedly would reflect felt security. Second, in the sociometer theory perspective on self-esteem (Leary, Tambor, Terdal, & Downs, 1995), the key construct is a person's expectations of how others will value and respond to the self (i.e., PR, *not* solely self-esteem, as thought by Leary).

In the risk regulation model derived from interdependence theory (Murray, Holmes, & Collins, 2006), the central construct is perceived regard or trust (i.e., a person's expectations about how a particular significant other will value and respond to the self). Finally, in the interpersonal schemas model (Baldwin, 1992) and relational self-theory perspective (Andersen & Chen, 2002), the important concept is the "self in relation to other" schema, including expected acceptance by a significant other—a concept surely closely overlapping with expected responsiveness (Reis et al., 2004).

Cameron and Holmes (2009) found that their new measure of perceived responsiveness correlated about 0.70 with the key constructs from all four major theories when they were measured in generalized form (i.e., for typical close others, suggesting very major theoretical agreement and overlap). Researchers need to replicate and extend these findings. However, if they are valid and reliable, much more effort must be devoted to examining common elements among different theories and integrating the insights derived by each.

Whereas a shared construct may be central to these varied theoretical approaches, there are, of course, very divergent perspectives on what the construct represents. For instance, both attachment and sociometer theories regard expectations about responsiveness as a generalized personality construct that will be applied across relational partners or significant others. The personality style is presumed to develop in early attachment experiences in the first case, with the addition of adolescent peer relations in the second. Risk regulation theory, on the other hand, contends that expected responsiveness from a *specific* partner is a function of both particular experiences with that partner and presumptions derived from personality influences, such as self-esteem or attachment style. Interpersonal schema perspectives adopt differing assumptions. Baldwin, Keelan, Fehr, Enns, and Koh-Rangarajoo (1996) suggested that even attachment styles can be specific to a partner but are likely to have personality roots. Andersen and Chen (2002) contend that schemas are very specific, tailored to each significant other in a person's life space.

These different perspectives on the responsiveness construct invite researchers to more thoroughly explore the relation of general and specific expectations to each other and to associated processes, as cogently argued by Collins, Guichard, Ford, and Feeney (2004). These are major theoretical issues with considerable importance within frameworks and exceptional practical significance. Are people essentially "fixed" in their expectations about other people from an early age? If they develop a close relationship where experiences seem to contradict their chronic assumptions, are they capable of largely supplanting the older attitude? Considering this issue raises some fascinating questions. For instance, when and why might an "old attitude" (e.g., attachment style) overturn a more recent one (i.e., partner-specific perceived responsiveness; Wilson, Lindsey, & Schooler, 2000)? Mikulincer and Shaver (2003) seem to suggest the interesting idea that basic attachment anxiety, the "old attitude," will prevail over a newer, positive expectations about a partner under stress or duress. On the other hand, the results of a study on women dealing with the stress of early parenthood seem to suggest that positive specific partner expectations buffered anxious women and resulted in their coping more effectively with the strains they faced (Rholes, Simpson, Campbell, & Grich, 2001). Finally, Murray and Holmes (2009) contend that positive specific partner attitudes also have an automatic component that shows itself most under stress or depletion. These challenging but rich issues deserve considerably more focus from researchers in relationship science.

In summary, interpersonal theories have much in common and need to celebrate their shared roots and not just emphasize their differences. Integration may spur new ideas within each framework.

Conclusion

Relationship science has made enormous progress in recent decades. In this chapter, I document four trends that I consider to be important to the progress in the discipline. However, the impressive set of chapters in this volume describes a wide variety of other major accomplishments that aptly demonstrate the growing maturity and breadth of research in the field. There is much to celebrate but also much to anticipate.

References

Andersen, S. M., & Chen, S. (2002). The relational self: An interpersonal social-cognitive theory. *Psychological Review, 109*, 619–645.

Baldwin, M. W. (1992). Relational schemas and the processing of social information. *Psychological Bulletin, 112*, 461–484.

Baldwin, M. W., Keelan, J., Fehr, B., Enns, V., & Koh-Rangarajoo, E. (1996). Social-cognitive conceptualization of attachment working models: Availability and accessibility effects. *Journal of Personality and Social Psychology, 71*, 94–109.

Bargh, J. A. (2007). *Social psychology and the unconscious: The automaticity of higher mental processes*. New York: Psychology Press.

Bargh, J. A., & Huang, J. Y. (2009). The selfish goal. In G. B. Moskowitz & H. Grant (Eds.), *The psychology of goals* (pp. 127–150). New York: Guilford Press.

Cameron, J. J., & Holmes, J. G. (2009). Searching for security: Perceived partner responsiveness as the core construct in theories of interpersonal relations. Unpublished manuscript. University of Waterloo.

Campbell, L., Simpson, J. A., Boldry, J., & Rubin, H. (2010). Trust, variability in relationship evaluations, and relationship processes. *Journal of Personality and Social Psychology, 99*, 14–31.

Cantor, N., & Kihlstrom, J. (1987). *Personality and social intelligence.* Englewood Cliffs, NJ: Prentice Hall.

Collins, N. L., Guichard, A. C., Ford, M. B., & Feeney, B. C. (2004). Working models of attachment: New developments and emerging themes. In W. S. Rholes & J. A. Simpson (Eds.), *Adult attachment: Theory, research, and clinical implications* (pp. 196–239). New York: Guilford.

Collins, N., & Read, S. J. (1994). Cognitivie representations of attachment: The structure and function of working models. In K. Bartholomew & D. Perlman (Eds.), *Advances in personal relationships: Attachment processes in adulthood* (Vol. 5, pp. 53–90). London: Kingsley.

Cook, W. L. (2000). Understanding attachment security in family context. *Journal of Personality and Social Psychology, 78*, 285–294.

Dijksterhuis, A., & Nordgren, L. F. (2006). A theory of unconscious thought. *Perspectives on Psychological Science, 1*, 95–109.

Downey, G., Freitas, A., Michaelis, B., & Khouri, H. (1998). The self-fulfilling prophecy in close relationships: Rejection sensitivity and rejection by romantic partners. *Journal of Personality and Social Psychology, 75*, 545–560.

Endler, N., & Hunt, J. (1969). Generalizability of contributions from sources of variance in the S-R inventories of anxiousness. *Journal of Personality, 37*, 1–24.

Fazio, R. H. (1986). How do attitudes guide behavior? In R. M. Sorrentino & E. T. Higgins (Eds.), *The handbook of motivation and cognition: Foundations of social behavior* (pp. 204–243). New York: Guilford Press.

Fitzsimons, G. M., & Finkel, E. J. (2010). Interpersonal influences on self-regulation. *Current Directions in Psychological Science, 19*, 101–105.

Fitzsimons, G. M., & Shah, J. (2008). How goal instrumentality shapes relationship evaluations. *Journal of Personality and Social Psychology, 95*, 319–337.

Fraley, C., Waller, N., & Brennan, K. (2000). An item response theory analysis of self-report measures of adult attachment. *Journal of Personality and Social Psychology, 78*, 350–365.

Gawronski, B., & Bodenhausen, G. V. (2006). Associative and propositional processes in evaluation: An integrative review of implicit and explicit attitude change. *Psychological Bulletin, 132*, 692–731.

Gable, S. L. (2005). Approach and avoidance social motives and goals. *Journal of Personality, 74*, 175–222.

Gilbert, D. T., & Malone, P. S. (1995). The correspondence bias. *Psychological Bulletin, 117*, 21–38.

Griffin, D. W., & Gonzalez, R. (2003). Models of dyadic social interaction. *Philosophical Transactions of the Royal Society: Biological Sciences, 358*, 573–582.

Hazan, C., & Shaver, P. (1987). Romantic love conceptualized as an attachment process. *Journal of Personality and Social Psychology, 52*, 511–524.

Holmes, J. G. (2000). Social relationships: The nature and function of relational schemas. *European Journal of Social Psychology, 30*, 447–496.

Holmes, J. G. (2002). Interpersonal expectations as the building blocks of social cognition: An interdependence theory perspective. *Personal Relationships, 9*, 1–26.

Holmes, J. G., & Cameron, J. (2005). An integrative review of theories of interpersonal cognition. In M. Baldwin (Ed.), *Interpersonal cognition* (pp. 415–447). New York: Guilford Press.

Holmes, J. G., & Murray, S. L. (2007). Felt security as a normative resource: Evidence for an elemental risk regulation system? *Psychological Inquiry, 18*, 163–168.

Holmes, J. G., & Rempel, J. K. (1989). Trust in close relationships. In C. Hendrick (Ed.), *Review of personality and social psychology: Close relationship*, (Vol. 10, pp. 187–219). Newbury Park, CA: Sage.

Kelley, H. H. (1979). *Personal relationships: Their structures and processes.* Hillsdale, NJ: Erlbaum.

Kelley, H. H., Holmes, J. G., Kerr, N., Reis, H., Rusbult, C., & Van Lange, P. A. (2003). *An atlas of interpersonal situations.* Cambridge, UK: Cambridge Press.

Kelley, H. H., & Thibaut, J. W. (1978). *Interpersonal relations: A theory of interdependence.* New York: Wiley.

Kenny, D. (1994). *Interpersonal perception: A social relations analysis.* New York: Guilford Press.

Kenny, D. A., & Cook, W. (1999). Partner effects in relationship research: Conceptual issues, analytic difficulties, and illustrations. *Personal Relationships, 6*, 433–448.

Kenny, D. A., Kashy, D., & Cook, W. (2006). *Dyadic data analysis.* New York: Guilford Press.

Leary, M. R., Tambor, E. S., Terdal, S. K., & Downs, D. L. (1995). Self-esteem as an interpersonal monitor: The sociometer hypothesis. *Journal of Personality and Social Psychology, 68*, 518–530.

Lewin, K. (1946). Behavior and development as a function of the total situation. In L. Carmichael (Ed.), *Manual of child psychology* (pp. 791–844). New York: Wiley.

McNulty, J. K., & Karney, B. R. (2004). Positive expectations in the early years of marriage: Should couples expect the best or brace for the worst? *Journal of Personality and Social Psychology, 86*, 729–743.

Mikulincer, M., Gillath, O., Halevy, V., Avihou, N., Avidan, S., & Eshkoli, N. (2001). Attachment theory and reactions to others' needs: Evidence that activation of the sense of attachment security promotes empathic responses. *Journal of Personality and Social Psychology, 81*, 1205–1224.

Mikulincer, M., Hirschberger, G., Nachmias, O., & Gillath, O. (2001). The affective components of the secure base schema: Affective priming with representations of attachment security. *Journal of Personality and Social Psychology, 81*, 305–321.

Mikulincer, M., & Shaver, P. R. (2001). Attachment theory and intergroup bias: Evidence that priming the secure base schema attenuates negative reactions to outgroups. *Journal of Personality and Social Psychology, 81*, 97–115.

Mikulincer, M., & Shaver, P. R. (2003). The attachment behavioral system in adulthood: Activation, psychodynamics, and interpersonal processes. In M. Zanna (Ed.), *Advances in experimental social psychology* (Vol. 35, pp. 52–153). New York: Academic Press.

Mischel, W., & Morf, C. C. (2003). The self as a psycho-social dynamic processing system: A meta-perspective on a century of the self in psychology. In M. R. Leary & J. P. Tangney (Eds.), *Handbook of self and identity* (pp. 15–46). New York: Guilford Press.

Mischel, W., & Shoda, Y. (1995). A cognitive-affective system theory of personality: Reconceptualizing situations, dispositions, dynamics, and invariance in personality structure. *Psychological Review, 102*, 246–268.

Morf, C. C., & Rhodewalt, F. (2001). Expanding the dynamic self-regulatory model of Narcissism: Research direcions for the future. *Psychological Inquiry, 12*, 243–251.

Muraven, M., & Baumeister, R. F. (2000). Self-regulation and depletion of limited resources: Does self-control resemble a muscle? *Psychological Bulletin, 126*, 247–259.

Murray, S. L., Aloni, M., Holmes, J. G., Derrick, J. L., Stinson, D. A., & Leder, S. (2009). Fostering partner dependence as trust insurance: The implicit contingencies of the exchange script in close relationships. *Journal of Personality and Social Psychology, 96*, 324–348.

Murray, S. L., Bellavia, G., Rose, P., & Griffin, D. (2003). Once hurt, twice hurtful: How perceived regard regulates daily marital interactions. *Journal of Personality and Social Psychology, 84*, 126–147.

Murray, S. L., Derrick, J., Leder, S., & Holmes, J. G. (2008). Balancing connectedness and self-protection goals in close relationships: A levels of processing perspective on risk regulation. *Journal of Personality and Social Psychology, 94*, 429–459.

Murray, S. L., & Holmes, J. G. (2009). The architecture of interdependent minds: A motivation-management theory of mutual responsiveness. *Psychological Review, 116*, 908–928.

Murray, S. L., Holmes, J. G., Aloni, M., Pinkus, R. T., Derrick, J. L., & Leder, S. (2009). Commitment insurance: Compensating for the autonomy costs of interdependence in close relationships. *Journal of Personality and Social Psychology, 97*, 256–278.

Murray, S. L., Holmes, J. G., & Collins, N. L. (2006). Optimizing assurance: The risk regulation system in relationships. *Psychological Bulletin, 132*, 641–666.

Murray, S. L., Holmes, J. G., & Griffin, D. W. (2000). Self-esteem and the quest for felt security: How perceived regard regulates attachment processes. *Journal of Personality and Social Psychology, 78*, 478–498.

Murray, S. L., Holmes, J. G., Griffin, D. W., Bellavia, G., & Rose, P. (2001). The mismeasure of love: How self-doubt contaminates relationship beliefs. *Personality and Social Psychology Bulletin, 27*, 423–436.

Murray, S. L., Holmes, J. G., MacDonald, G., & Ellsworth, P. (1998). Through the looking glass darkly? When self-doubts turn into relationship insecurities. *Journal of Personality and Social Psychology, 75*, 1459–1480.

Murray, S. L., Holmes, J. G., & Pinkus, R. T. (2010). A smart unconscious? Procedural origins of automatic partner attitudes in marriage. *Journal of Experimental Social Psychology, 46*, 650–656.

Murray, S.L., Rose, P., Bellavia, G., Holmes, J., & Kusche, A. (2002). When rejection stings: How self-esteem constrains relationship-enhancement processes. *Journal of Personality and Social Psychology, 83*, 556–573.

Olson, M. A., & Fazio, R. H. (2009). Implicit and explicit measures of attitudes: The perspective of the MODE model. In R. E. Petty, R. H. Fazio, & P. Brinol (Eds.), *Attitudes: Insights from the new implicit measures* (pp. 19–63). New York: Psychology Press.

Overall, N. C., & Sibley, C. G. (2009). When rejection-sensitivity matters: Regulating dependence within daily interactions with family and friends. *Personality and Social Psychology Bulletin, 35*, 1057–1070.

Pierce, T., & Lydon, J. (1998). Priming relational schemas: Effects of contextually activated and chronically accessible interpersonal expectations on responses to a stressful event. *Journal of Personality and Social Psychology, 75*, 1441–1448.

Reis, H. T. (2008). Reinvigorating the concept of situation in social psychology. *Personality and Social Psychology Review, 12*, 311–329.

Reis, H. T., Clark, M. S., & Holmes, J. G. (2004). Perceived partner responsiveness as an organizing construct in the study of intimacy and closeness. In D. Mashek & A. P. Aron (Eds.), *Handbook of closeness and intimacy* (pp. 201–225). Mahwah, NJ: Lawrence Erlbaum.

Reis, H., & Holmes, J. G. (in press). Perspectives on the situation. In K. Deaux & M. Snyder (Eds.), *The Oxford handbook of personality and social psychology*. Oxford, UK: Oxford University Press.

Reis, H. T., & Judd, C. M. (Eds.). (2000). *Handbook of research methods in social psychology*. New York: Cambridge University Press.

Rholes, S., Simpson, J., Campbell, L., & Grich, J. (2001). Adult attachment and the transition to parenthood. *Journal of Personality and Social Psychology, 81*, 421–436.

Ross, L., & Nisbett R. E. (1991). *The person and the situation: Perspectives of social psychology*. Philadelphia: Temple University Press.

Rusbult, C. E., & Van Lange, P. A. M. (2003). Interdependence, interaction, and relationships. *Annual Review of Psychology, 54*, 351–375.

Simpson, J. A., Ickes, W., & Blackstone, T. (1995). When the head protects the heart: Empathic accuracy in dating relationships. *Journal of Personality and Social Psychology, 69*, 629–641.

Simpson, J. A., Rholes, W. S., & Phillips, D. (1996). Conflict in close relationships: An attachment perspective. *Journal of Personality and Social Psychology, 71*, 899–914.

Strack, F., & Deutsch, R. (2004). Reflective and impulsive determinants of social behavior. *Personality and Social Psychology Review, 8*, 220–247.

Thibaut, J. W., & Kelley, H. H. (1959). *The social psychology of groups*. New York: Wiley.

Wilson, T. D., Lindsey, S., & Schooler, T. Y. (2000). A dual model of attitudes. *Psychological Review, 107*, 101–126.

Author Index

A

B

C

D

E

F

G

H

I

J

K

N

O

P

R

S

T

U

V

W

Y

Z

Subject Index

B

C

D

E

I

J

L

M